THE NEW CORPORATE CLIMATE LEADERSHIP

This book provides a comprehensive treatment of the role of the private sector in accelerating the transition to a low-carbon, climate-resilient, and inclusive world.

In the lead up to and since the historic Paris Agreement on climate change, more than 6,000 companies from 120 countries representing more than $36.5 trillion in revenue have made climate commitments. Examining this trend, *The New Corporate Climate Leadership* provides a clear synthesis of the relationship between the real economy and climate change and offers a state-of-the-art assessment of corporate initiatives that focus on greenhouse gas emissions reductions and the management of climate risk through enhanced resilience. It debates the relative merits of incremental and sequenced ambition versus radical systems change – including a critique of the prevailing capitalist approach to climate change – and provides an actionable guide to skills development for change-makers in the shift toward a low-carbon world. Drawing on perspectives from leading thinkers inside the private sector, across government, and within civil society to truly interrogate the scale, scope, and speed of progress, this book provides a clear vision for what the next generation of corporate climate leadership should look like.

Optimistic in tone, this book will be of great interest to students, scholars, and practitioners of climate change and sustainable business.

Edward Cameron is an independent consultant serving governments, companies, philanthropies, international organizations, and non-profits.

Emilie Prattico holds a PhD in philosophy from Northwestern University, a Master's degree from HEC Paris, and a BA degree in Philosophy and Theology from the University of Oxford. She has worked at the intersection of the public and private sector on sustainability and climate for ten years and currently focuses on consulting for private sector companies.

Routledge Research in Sustainability and Business

Corporate Responsibility and Sustainable Development
Exploring the Nexus of Private and Public Interests
Edited by Lez Rayman-Bacchus and Philip R. Walsh

Sustainable Entrepreneurship and Social Innovation
Edited by Katerina Nicolopoulou, Mine Karatas-Ozkan, Frank Janssen and John Jermier

Corporate Social Responsibility, Human Rights and the Law
Stéphanie Bijlmakers

Corporate Sustainability
The Next Steps Towards a Sustainable World
Jan Jaap Bouma and Teun Wolters

Strategic Corporate Social Responsibility in Malaysia
Edited by Rusnah Muhamad and Noor Akma Mohd Salleh

Renewable Energy Enterprises in Emerging Markets
Strategic and Operational Challenges
Cle-Anne Gabriel

Accountability, Philosophy and the Natural Environment
Glen Lehman

Corporate Social Responsibility in the Arctic
The New Frontiers of Business, Management, and Enterprise
Gisele M. Arruda and Lara Johannsdottir

The New Corporate Climate Leadership
Edward Cameron and Emilie Prattico

For more information about this series, please visit: www.routledge.com/Routledge-Research-in-Sustainability-and-Business/book-series/RRSB

THE NEW CORPORATE CLIMATE LEADERSHIP

Edward Cameron and Emilie Prattico

First published 2022
by Routledge
2 Park Square, Milton Park, Abingdon, Oxon OX14 4RN

and by Routledge
605 Third Avenue, New York, NY 10158

Routledge is an imprint of the Taylor & Francis Group, an informa business

© 2022 Edward Cameron and Emilie Prattico

The right of Edward Cameron and Emilie Prattico to be identified as authors of this work has been asserted in accordance with sections 77 and 78 of the Copyright, Designs and Patents Act 1988.

All rights reserved. No part of this book may be reprinted or reproduced or utilised in any form or by any electronic, mechanical, or other means, now known or hereafter invented, including photocopying and recording, or in any information storage or retrieval system, without permission in writing from the publishers.

Trademark notice: Product or corporate names may be trademarks or registered trademarks, and are used only for identification and explanation without intent to infringe.

British Library Cataloguing-in-Publication Data
A catalogue record for this book is available from the British Library

Library of Congress Cataloging-in-Publication Data
A catalog record has been requested for this book

ISBN: 978-0-367-45883-6 (hbk)
ISBN: 978-0-367-45885-0 (pbk)
ISBN: 978-1-003-02594-8 (ebk)

DOI: 10.4324/9781003025948

Typeset in Bembo
by KnowledgeWorks Global Ltd.

CONTENTS

Acknowledgments vii

Introduction 1

PART I
Climate risks, response, and rewards for the 21st-century company

1 Climate change: The highest impact risk for global business 21

2 Overcoming systemic barriers: Shifting to a new climate economy 37

3 The greatest economic opportunity in history: Building a new climate economy 63

PART II
The current corporate climate leadership

4 Building the climate-compatible economy: Strategies for corporate decarbonization and resilience 89

5 Current corporate leadership: How commitments and inconsistencies shape the private sector response to climate crisis 110

PART III
The new corporate climate leadership

6 The innovation agenda: Reimagining the
 climate-compatible business   131

7 Financial rewards: Rethinking assets and liabilities in
 the context of climate change   152

8 Corporate governance, purpose, and stakeholders:
 Transforming companies to transform the economy   168

9 From evolution to revolution: Civil society and rethinking
 business in a new climate economy   185

 Conclusion: Vision 2030: building a low-carbon,
 climate-resilient, and inclusive world   210

Index   *217*

ACKNOWLEDGMENTS

This book was written by two authors who yearn for a just and sustainable world, appreciate the enormity of the task at hand, and are inspired in their work by countless individuals across the world. This work is a note of gratitude to all the ordinary and some not-so-ordinary individuals who continue to enhance our understanding of climate science; shape public policy to protect the environment and advance human rights; advocate for justice; rethink the role of companies to protect people and planet as well as pursue profit; and empower the next generation of leaders to be agents of positive change in the world.

As longstanding participants in global efforts to avert climate crisis and create a shared prosperity, we are grateful for the many organizations we have served, and the countless colleagues we have had, who dedicate their professional lives to a sincere attempt at building a more inclusive world. This book is founded on their expertise, sourced from their years of diligent research, and results from their willingness to educate and mentor the next generation of climate leaders. In particular, we would like to thank the professionals serving at hundreds of the world's leading companies who were willing to share their sustainability struggles as well as their successes with great integrity to help us paint a picture of how far corporate climate leadership has come, and how far it still needs to travel over the coming decade.

The authors have many friends, teachers, allies, partners, and colleagues to thank for helping them reach the point where they felt inspired and able to write this book, but first thanks go to each other. We both recognize that whether it was in the years of practicing climate work together before the idea of this book even came about or through the months of co-writing this volume that each other's encouragement, criticism, friendship, shared skepticism, and humor were essential to bringing this book to our audience.

Edward would like to thank his wife Carina Bachofen, a leader in the humanitarian community, for teaching him about vulnerability and resilience and for insisting that the needs of frontline communities should always be the highest goal of climate and development work. Edward would also like to thank the many former colleagues at the World Resources Institute, BSR, and We Mean Business for insisting on rigor, building an understanding that change has to happen in the real economy, not in conference rooms, and for pioneering the corporate climate leadership that provides the foundations for climate-compatible development over the coming decades. Finally, Edward would like to dedicate this work to his young sons – Robert and James – who will one day be called to invest their own professional energy in building a just and sustainable world.

Emilie would like to thank a few true leaders and visionaries, with whome she has had the privilege of weaving dialogues with over the years: Gil Friend, Dimitri Caudrelier, Alison Taylor, Mathieu Menegaux, Tara Norton, Guy Morgan, and Ariane Vitou. Their guidance in all things climate, CSR, corporate leadership, and finding one's own path in a slippery field has proved invaluable both personally and professionally. For his warm welcome at the Institute of Futures Studies when this book was just an idea and a plan, and for his long-lasting friendship, Gustaf Arrhenius. For their unflinching support, of the kind that sheds light on dark corners, I would like to thank Stuart Pickles, Claude Grunitzky, Issa Kohler-Hausmann, Ashwin Jacob, and Deborah Allen Rogers, and a special group of classmates from the HEC Paris Masters in Sustainable Development (class of 2012), dedicated not only to the issues presented in this book but to achieving transformation with integrity and joy. Finally, her deepest thanks go to those whose support cannot be expressed in words: her parents, Brenda and Antonino, who witnessed most of this book come to life in their home during the COVID-19 pandemic; her sisters, Jane and Anne-Marie; her partner, Austin Taylor; and her daughter, who, in large part, has inspired not only this book but all her efforts to bring about a better world.

Together, Edward and Emilie would like to thank the following colleagues for offering their time and expertise in reviewing this work: Myriam Coneim, Roop Singh, Carina Bachofen, and Michale Philipp. We would also like to thank the blind reviewer provided by the publisher for offering considerable insights and improvements.

Last but by no means least, Edward and Emilie would like to express their deep gratitude to the colleagues at Routledge who commissioned, guided, and ultimately produced this book. This includes Annabelle Harris, Matthew Shobbrook, Cathy Hurren, and Nancy Rebecca.

INTRODUCTION

It is time for a *new* corporate climate leadership. Most companies are new to the cause of climate action, and even those early pioneers of ambition are finding that leadership is dynamic: today's leaders can quickly become tomorrow's laggards. Just as the impacts of climate change keep accelerating, so must the pace at which we address them. This kind of exponential pace of change should not seem new to companies. We saw it in the so-called digital revolution, and it is indeed a feature of most breakthrough transformations. But in the case of climate change, failing to adopt a new form of leadership is far more than a business risk.

The human toll of climate change is enormous. The Lancet Commission estimates that fossil fuel pollution caused an estimated 9 million premature deaths in 2015.[1] Natural disasters doubled in the period 2000–19 when compared to 1980–99. Climate-related disasters account for the difference, rising from 3,600 to 6,700 and affecting 4.2 billion people across the globe.[2] In 2017, climate-related disasters caused acute food insecurity for approximately 39 million people across 23 countries. And the combination of economic deprivation, ecological marginalization, and involuntary displacement are increasingly causing "deprivation conflicts" such as civil strife and insurgency and group identity conflicts in fragile states.

The economic impacts are also significant and growing. The World Economic Forum (WEF) has consistently ranked climate-related events such as extreme weather events, increased incidence of drought and flood; and ecosystem degradation as the highest material risks to global business in both likelihood and impact.[3] The global cost could be as high as $24 trillion by 2030,[4] and the Financial Stability Board, which represents ministries of finance and central banks from across the G20, estimates the total stock of manageable assets at risk to be $43 trillion between now and the end of the century.[5] In the United States, climate-related disaster losses are growing at a rate of 6% a year and costing an

DOI: 10.4324/9781003025948-1

2 Introduction

average of $100 billion annually.[6] In 2018, Hurricanes Florence and Michael were just two of 14 "billion-dollar disasters" in the United States, while the 2017 Atlantic hurricane season was the costliest on record with over $290 billion in damages from Harvey and Irma alone.[7] BlackRock calculates a 275% increase in major hurricane risk by 2050, raising the stakes still further.[8] A range of global brands including Verizon and Hewlett-Packard have suffered multi-billion dollar losses from climate-induced supply chain disruption. In 2019, PG&E announced the biggest utility bankruptcy in US history in response to $30 billion in fire liabilities associated with the wildfires in Northern California in 2017 and 2018. They were forced into an $11 billion settlement, and their share price collapsed by 90%.[9] The message is clear; the cost of climate inaction is increasing for the wider economy and for individual companies.

In January 2021, a group of the world's leading natural scientists published a short and stark overview of the interrelated global risks facing humanity over the coming decades. The climate crisis featured prominently in their assessment. Writing candidly, they suggested "the choice before us is between exiting overshoot by design or disaster – because exiting overshoot is inevitable one way or another."[10]

This book makes a case for exiting overshoot by design, led by a new corporate climate leadership committed to decarbonization, resilience, and a shared prosperity.

Motivations and methods

In 1965, Lyndon Baines Johnson occupied the White House as the 36th President of the United States. In November of that year the President's science advisory committee published a report that foresaw the coming climate crisis. Written by the most prominent climate scientists of the age – Roger Revelle and Charles David Keeling amongst others, the report warned that "man is unwittingly conducting a fast geophysical experiment which will almost certainly cause significant changes in temperature." In the foreword to the Report, the President wrote, "Ours is a nation of affluence. But the technology that has permitted our affluence spews out vast quantities of wastes and spent products that pollute our air, poison our waters, and even impair our ability to feed ourselves."[11]

We have decided to write this book because we cannot afford to sit on what we know for another 55 years. In our careers we have chronicled the journey of climate change from silence to salience, from the margins to the mainstream. We are aware of the unprecedented level of public policy commitment to decarbonization, personified by the landmark Paris Agreement and the 189 national climate action plans that have emerged to honor the Agreement. We are further aware of the historic level of commitment from non-state actors including provinces, cities, municipalities, and the six thousand multinational companies, representing over half the global economy, who have pledged to reduce their greenhouse gas emissions. However, we are also aware that we are failing to meet the moment. The past decade has seen a rapid rise in greenhouse gas emissions, and what was projected as the climate change of tomorrow has become the natural disasters we face already today. We need to transform over the coming decade, and that must start with corporations and the real economy. We therefore offer this book as a means for companies to understand climate risk and embark on a new decade of dynamic climate leadership.

Our work is rooted in "critical theory framework," as this research was conducted in order to catalyze a positive influence on society, and particularly on global corporations and those who will lead them.[12] We did not set out to be value free or neutral in the work – our goal is to outline the risks of inaction and the rewards of leadership in order that the private sector accelerates the transition to a low-carbon, resilient and inclusive economy.

We used a combination of research methods to build the evidence base and make our argument including:

- Documentary and Literature Analysis. A large volume of peer-reviewed and published scientific articles from respected academic journals, international organizations, research institutions, universities, the family of UN agencies, global think tanks, non-profits, and companies. On issues related to climate science we stuck rigorously to the findings of the Intergovernmental Panel on Climate Change (IPCC), the gold standard for assessing and communicating the scientific basis of climate change, its impacts, and future risks.
- Practitioner Interviewing: Given the focus of this work is on understanding how the private sector is responding to climate change while advising on how it should respond in future through a dynamic form of leadership; it was important to study those working in the private sector and to engage in semi-structured interviews with thinkers inside companies. We followed the practice known as "snowball" or "referral sampling," whereby we began with contacts in our own immediate circle and asked them to recommend additional interviewees. These interviews were conducted anonymously as those being interviewed needed to be frank about shortcomings as well as the strengths of their employers' approach to climate change.
- Participant observation: Together the authors have more than 40 years working at the interface of climate change, development and the private sector and have drawn on that extensive experience in writing this book.

The lost decade

Ta-Nehisi Coates begins his masterful assessment of what the Obama Presidency meant for race relations in the United States with the words: "this story began, as all writing must, in failure."[13] Unfortunately, this story begins with failure as well. The United Nations Environment Programme (UNEP) has described "a decade lost" in a recent assessment of climate policies. Describing the findings as "sobering," the analysis reveals that global greenhouse gas (GHG) emissions in

2018 were almost exactly at the level of emissions projected for 2020 under the "business as usual" or "no policy" scenarios used in the Emissions Gap Report of 2011. In other words, there has been no real change in the global emissions pathway over the last decade.[14] This is remarkable given that 140 countries endorsed the Copenhagen Accord in 2009, with 85 of them pledging to reduce their emissions through national policies. The unprecedented Paris Agreement of 2015 was subsequently adopted by 195 countries, 184 of which have so-called nationally determined contributions (or national climate plans) designed to limit global warming to well below 2°C.[15] The analysis goes on to report that because GHG emissions continue to grow, governments must now triple the level of ambition reflected in their current and planned climate policies to get on track toward limiting warming to below 2°C, while at least a fivefold increase is needed to align global climate action and emissions with limiting warming to 1.5°C by the end of this century.[14]

The "slow violence of unsustainability"[16]

The results of this lost decade are stark. Long before COVID-19, the world was already suffering a series of silent public health emergencies with a clear link to the climate crisis. This included food insecurity, fossil fuel pollution, and mortality from extreme events.

Food security requires a consistent, available, and accessible source of food of sufficient quantity and quality to satisfy nutritional needs. The World Food Programme (WFP) defines acute food insecurity as the lack of secure access to sufficient amounts of safe and nutritious food for normal human growth and development and an active and healthy life, while chronic food Insecurity is a long-term or persistent inability to meet dietary energy requirements.[17]

Separate research conducted by the WFP, Food and Agriculture Organization (FAO), and UNICEF during the past two years reveal a world suffering from substantial food insecurity. WFP estimates that 135 million people were in a state of acute food insecurity in 2019, with an additional 183 million people at risk of slipping into acute food insecurity. This included an estimated 75 million children under five years old with limited access to nutrition.[18] FAO has concluded that more than 820 million people suffer from hunger.[19] What's more, when the organization broadens its horizon to consider those suffering even moderate levels of food insecurity, the FAO concludes that over 2 billion people do not have regular access to safe, nutritious, and sufficient food, including 8% of the population in Northern America and Europe.[19] At the same time, changes in consumption patterns have contributed to about two billion adults now being overweight or obese,[20] contributing to 4 million deaths globally every year.[19] Moreover, 25–30% of total food produced is lost or wasted.[20]

In a report for UNICEF, Development Initiatives found that slow progress on malnutrition could cost society up to $3.5 trillion per year, with weight and obesity issues alone costing $500 billion per year.[21] According to UNICEF, an estimated 6.3 million children under 15 years of age died in 2017, mostly of preventable causes such as lack of access to water, sanitation, proper nutrition, or basic health services. The majority of these deaths – 5.4 million – occur in the first five years of life due to preventable or treatable causes such as complications during birth, pneumonia, diarrhea, neonatal sepsis, and malaria.[22]

The Lancet Commission estimates that fossil fuel pollution is the largest environmental cause of disease and premature death in the world, responsible for an estimated 9 million premature deaths in 2015 or 16% of all deaths worldwide. People in more than 90% of cities breathe polluted air that is toxic to their cardiovascular and respiratory health.[23] Asia accounts for more than half of the 9 million pollution-related fatalities recorded in 2015. India's 2.5 million deaths are by far the biggest share, with pollution from burning coal being the principal cause.[24]

The health implications include an increase in transmission of vector- and water-borne diseases. The number of people getting diseases transmitted by mosquitos, ticks, and fleas has more than tripled in the United States since 2004. Between 2004 and 2016, about 643,000 cases of insect-borne illnesses were reported to the Centers for Disease Control (CDC); rising from 27,000 per year in 2004 to 96,000 by 2016. The CDC estimates that 300,000 Americans get Lyme disease every year but only about 35,000 are reported. The causes are ticks thriving in areas that were previously too cold for them and hot spells triggering outbreaks of mosquito-borne diseases.[25]

Extreme events have also taken a human toll. The International Federation of Red Cross and Red Crescent Societies (IFRC) calculates that between 1996 and 2015 extreme weather killed 528,000 people worldwide, while 92% of natural hazards were climate related, and the total damage came to $3.08 trillion.[26] According to the World Meteorological Organization (WMO), in 2018 weather and climate events impacted nearly 62 million people, with floods affecting more than 35 million people and over 9 million people affected by drought.[27] Extreme weather events caused a total of $306 billion in damages across the United States in 2017, making the year the most expensive year on record for natural disasters. There were 16 disasters that caused more than $1 billion in 2017.[28] In California, a half-decade-long drought was followed by winter rains in 2016 and 2017, which encouraged rapid plant growth. Then, intense heat dried out the land and triple-digit heat fueled even greater fires. Climate change is responsible for about half the additional drying that researchers have found since the 1970s, resulting in a doubling of the area forest fires have consumed since 1984. Climate change may also increase lightning strikes, which are a major source of wildfires, and generate the high winds that can drive big blazes. Meanwhile, earlier springtime melting means that the land has more time to dry out over the warmer months.[29]

The Atlantic Meridional Overturning Circulation (AMOC) – one of the Earth's major ocean circulation systems – redistributes heat on our planet and has a major impact on climate. New research suggests there is evidence the AMOC is slowing down in response to anthropogenic global warming and that the

AMOC is currently in its weakest state for more than 1,000 years. If these trends hold or accelerate, the consequences could include faster sea level rise along parts of the Eastern United States and parts of Europe, and stronger hurricanes in the Southeastern United States.[30]

In addition, the combination of poverty, hunger, health stressors and conflict forced an estimated 79 million people to be displaced globally as of mid-2019 – 44 million of them were internally displaced while 20 million were refugees under theUN's protection. More than half of these refugees were hosted in countries with high numbers of acutely food-insecure people.[17]

The decisive decade

In this new decade we must avoid unmanageable climate change by pursuing rapid and aggressive decarbonization; manage unavoidable climate change by building resilience in the face of existing climate impacts; and create a shared prosperity where the most vulnerable are included in just transition to a new economy.

Net-zero greenhouse gas emissions by 2050

To avoid unmanageable climate change and safeguard both human and natural systems, we must hold global mean temperature rises to less than 1.5°C. According to the IPCC in model pathways with no or limited overshoot of 1.5°C, global net anthropogenic CO_2 emissions need to decline by about 45% from 2010 levels by 2030 (40–60%, reaching net-zero emissions by 2050.[31] Net-zero is the point when greenhouse gas (GHG) emissions from human activities at source are balanced globally by GHG removals. Reducing emissions at source involves a large range of strategies, focused principally on energy, transportation, and land use. This includes improving energy efficiency, shifting toward renewable energy, moving toward plant-based diets, reducing food wastes, and expanding low-carbon mobility options for people, goods, and services. In scenarios limiting warming to 1.5°C, carbon dioxide (CO_2) needs to reach net-zero around 2050, while total GHG emissions must reach net-zero by 2068.[32]

Achieving this goal is a massive political and financial undertaking, and it will require greater political will and the level of public consensus that we have not previously seen. However, all of the innovations and technology necessary to achieve this goal are already available to us. Analysis from the UNEP reveals that this is within reach. Investments in just six areas – solar energy, wind energy, efficient appliances, efficient cars, afforestation, and halting deforestation – would build a pathway to net-zero.[33] Project Drawdown – an unprecedented study of emissions reductions measures to stop GHGs in the atmosphere climbing and start to steadily decline – has identified 80 solutions that are available today that could deliver emissions reductions in these six areas while driving economic development, job creation, the democratization of energy access, and food sovereignty.[34]

Similarly, the IPCC has identified 60 so-called response options that can reduce GHG emissions from land use, enhance adaptive capacity, minimize land degradation, and increase food security. For example, reducing food waste by 50% would generate net emissions reductions in the range of 20–30% of total food-sourced GHGs. Shifting to diets that are lower in emissions-intensive foods like beef delivers a mitigation potential equivalent to the emissions of the whole of North America.[20]

Carbon removal can be achieved by enhancing carbon sinks, using natural strategies like afforestation, agricultural soil management, and marine protected areas and expanding carbon sequestration such as direct air capture and storage (DACS) technology, carbon capture and storage, and enhanced mineralization. For example, expanding, restoring, and managing forests can leverage the power of photosynthesis, converting carbon dioxide in the air into carbon stored in wood and soils. The World Resources Institute (WRI) estimates that the carbon-removal potential from forests in the United States alone is equivalent to all annual emissions from the U.S. agricultural sector, is inexpensive compared to other carbon removal options, and can deliver co-benefits such as cleaner water and air.[35]

Resilient and inclusive focused on vulnerable populations and gender equality

We must also invest in resilience, meaning our capacity to anticipate, avoid, accommodate, and recover from shocks. This, in turn, requires investment in six capital assets – human, social, natural, physical, financial, and political – with a focus on vulnerable populations who experience intersecting political, economic, legal, social, and cultural inequalities that amplify their risk; and those who require a just transition to move away from high-carbon pathways.

To be truly resilient, we must build an inclusive economy to advance the realization of human rights – all rights including civil, political, economic, and social. The body of international human rights law, dating back to the Universal Declaration of Human Rights in 1948, provides the enabling conditions to achieve this goal. We must double down on the so-called "procedural rights" that facilitate advocacy and accountability and are critical to building climate resilience. These include:

- Access to information provides opportunities for individuals to understand and develop agency to mitigate the impacts of climate change. Access to information is anchored in the human rights that guarantee access to education, free assembly and association, and free opinion and expression. It allows otherwise marginalized and vulnerable groups to gain insight into what inputs are shaping policy-making and to remain informed on how policy is being implemented and enforced.

- Access to decision-making is anchored in rights to shape governing political systems, access a free press to be informed by and to inform decision-making, access fair and open public administrative and judicial hearings, and access decision-making to promote participatory policy-making. Importantly, vulnerable populations should be counted among those shaping climate change interventions.
- Access to justice through various tribunals provides the scope for dispute settlement and redress when policies are poorly conceived, processes are not respected, or outcomes are harmful. Access to justice enables communities to hold the public and private sector accountable for failures to build resilience in a manner that is proportional.[36]

The focus on rights should include an explicit commitment to address gender inequalities. Measures to improve gender equality should include improving women's access to financial services including savings, insurance, and credit, and expanding ownership of productive assets such as property, farms, and inheritance. Efforts should be made to expand labor relations law and social safeguards including minimum wage, maternity benefits, and old age pensions to women concentrated in informal and unprotected work. Greater emphasis should be placed on ensuring that women are neither underpaid nor unpaid. And expanding educational opportunities for women, including access to agricultural extension and rural advisory services, should be a priority. Educating girls, often referred to as the single best investment for development, leads to better employment opportunities for those girls in adulthood and to those adults raising healthier, more educated children. Involving women in community-level decision-making processes tends to produce an increased focus on public goods, such as education, as well as water and sanitation services. Closing the gender gap in agriculture would generate significant gains for the agriculture sector and for society. If women had the same access to productive resources as men, they could increase yields on their farms by 20–30%. This could raise total agricultural output in developing countries by 2.5–4%, which could, in turn, reduce the number of hungry people in the world by 12–17%. Countries where women lack any right to own land have on average 60% more malnourished children.[37]

A resilient economy also provides for a "just transition" for those individuals, households, and communities currently invested in high-carbon development as they must also be invited to share in the prosperity of the new economy. The International Labour Organization (ILO) has predicted that employment will be affected as climate policies and corporate leadership define a low-carbon economy. The expansion of greener products, services, and infrastructure will translate into higher labor demand across many sectors of the economy and associated new jobs. Some existing jobs will be substituted as we shift from internal combustion engine manufacturing to electric vehicle production, or from coal to solar power. This shift may be particularly difficult for low-skilled workers as they will need to retrain to compete for new jobs. Moreover, locations losing employment may not be the same places to benefit from the new jobs, and so geographic disparities may increase. Certain jobs or sectors may be eliminated – either phased out or massively reduced in numbers. This is particularly acute for fossil fuel producers and carbon-intensive industries or practices. Many existing trade jobs will be transformed and redefined as day-to-day workplace practices, skill sets, work methods, and job profiles are greened. For example, automobile workers will produce more electric cars, while farmers will engage in sustainable land management, address food loss and waste, and respond to changing dietary habits.[38] A resilient economy will therefore seek to strengthen social dialogue, social protection systems, vocational training, and transfer payments and investments in communities suffering deindustrialization and dislocation.

Issues of resilience and human rights are brought into stark relief by the current COVID-19 crisis. One of the key lessons, as we seek to learn from this experience, ought to be the need for expanded investment in healthcare systems; expanded access to healthcare; reduced healthcare costs; a commitment to democratizing preventative measures; and strengthening health-related human rights including sanitation, food, housing, working conditions, and reproductive rights.

The international framework to support this is already there. Article 25 of the Universal Declaration of Human Rights states that "everyone has the right to a standard of living adequate for the health and well-being of himself and of his family, including food, clothing, housing and medical care…"[39] Article 12 of the International Covenant on Economic, Social and Cultural Rights recognizes the right of everyone to the enjoyment of the highest attainable standard of physical and mental health.[40] According to the Office for the United Nations Commissioner for Human Rights (OHCHR) the right to health includes a wide range of factors, including access to safe drinking water and adequate sanitation; safe food; adequate nutrition and housing; healthy working and environmental conditions; health-related education and information; the right to a system of health protection providing equality of access for everyone; and access to essential medicines.[41]

Going forward, overcoming the challenges created by the lost decade will now require additional annual energy-related investments of between $830 billion and $2.4 trillion – about 2.5% of the world GDP (IPCC 2018). Mobilizing finance at this scale and transforming the real economy will require unprecedented leadership from the world's major corporations.

The current leadership

Thirty-three years after James Hansen delivered presented models to the US congress, saying he was "99 percent sure" that global warming was upon us, it is fair to say corporate leadership has emerged, and it is fair to say, too, that it has emerged against all odds. At the same time as science entered the hemicycle, some companies in high-emitting sectors were spending billions of dollars to sway public and political opinion against it. This set the course for an uphill battle against a range of skeptics, from deniers to obstructionists, for anyone who wanted to make headway to avert the worst impacts of climate change – corporate or not.

And yet companies have made significant progress in at least two ways: in shifting mindsets so that their role in climate action is no longer in question and in adopting measures to reduce their environmental footprint including their climate footprint. There is hardly any comparison between efforts to galvanize corporate action in the early 1990s, a period where most Americans and Canadians still believed climate change was due to space exploration,[42] and today, where a third of the UK FTSE has pledged to reach net-zero carbon emissions by 2050.[43]

Progress in terms of the number of companies taking a stand and measures is indeed undeniable. As of April 2021, over six thousand companies and investors, representing $36 trillion in revenue, have committed to climate action. These companies, from 120 countries, make up half the global economy.[44] These pioneering companies are compelling their suppliers to disclose climate risk and set emissions reductions as a business condition.[45]

This evolution has not been linear. The run-up to the landmark Paris Agreement saw unprecedented mobilization of private sector companies, who rightly predicted high expectations from civil society during the 2015 conference, and who saw it as an opportunity to showcase contribution to, if not leadership on, the goals of the Agreement. Through an effect of critical mass, climate action has since then become an almost unavoidable element of any corporate social responsibility (CSR) strategy or new business strategy announcement. There is no doubt, then that the role the private sector sees itself as having – and the responsibilities it is willing to take publicly and vis-à-vis shareholders – in the context of climate action is unprecedented.

That we are talking here about leadership is quite clear from the fact this sort of climate action emerged in a context where any environmental action was seen as a cost, a reputational risk of being accused of greenwashing, and with a potential political dimension that companies have traditionally distanced themselves from in their open communications. The meaningful kind of climate action we are witnessing now might have taken courage for first movers and certainly the willingness to take a stand – and risks – with major stakeholders.

The question, however, whether this kind of leadership is fit for the purpose of facing today's climate crisis is a live one. As we have outlined above, the kind of climate action we need goes far beyond the confines of carbon accounting. If we could imagine all companies in the world reducing their emissions in line with the IPPC's recommendations so that we may remain below the 1.5°C threshold, we would not have addressed the impacts of climate change exhaustively or adequately. If, in this hypothetical scenario, no changes were made to current inequalities in victims of climate impacts, if not consideration was taken for those who may have lost out economically or socially in the transition to a new low-carbon economy, if those displaced by climate impacts remain treated as pariahs, then we will not have, in fact, addressed climate change. We will have reduced our emissions to avert some of its worst impacts. While this is a necessary condition to addressing climate change, it is not sufficient.

And yet today, the focus, in climate leadership, is overwhelmingly on emissions reductions. The growth of the Net Zero movement in the private sector demonstrates just this. It is essential that this become the "new normal," as the UN Secretary-General called for in February 2021,[46] there is no doubt about that. But focusing exclusively on emissions reductions without a view to transforming the systems that favored unbridled emissions over the past century, without regard to those who will be most affected by the impacts of climate change, or without caution for the future we are building together, does not constitute the kind of leadership we need today.

The new corporate climate leadership

What we aim to present here, then, is a pathway to this kind of leadership. This is not a guide to setting a net-zero carbon emissions strategy. It is a guide to erecting the kind of leadership that will take us into a world where climate risk is not merely managed but rightly diagnosed and transformed into a blueprint for a more equitable world. It is a kind of leadership that understands and even relies on the imperatives of business to advance these goals, one that seizes economic opportunities in favor of the transition to the new climate economy with a longer-term vision and a new definition of value creation.

Based on our understanding of current climate action and of the urgency of transformative, we have outlined a pathway to the kind of climate leadership that has a chance at tackling the climate predicament of this day. It too may one day become obsolete and will need to be surpassed by yet a different model. But we are certain that if we do not collectively push for – and implement ourselves where we can – a new climate leadership that is systemic in breadth, rights-orientated, and radically transformative of existing models of value creation, our chances of building the world we want are close to zero.

Overview of chapters

Part I, Chapter 1: Climate change: the highest impact risk for global business

The Financial Stability Board (FSB) has identified climate change as a significant risk to the economy. The WEF has consistently ranked climate change as the highest material risk to global business in both likelihood and impact. The global cost to business could be as high as $24 trillion by 2030 and $43 trillion by the end of the century. Properly diagnosing and responding to climate risk is becoming an essential component of broader enterprise risk management, vital to securing competitive advantage. And yet, most companies are failing to properly understand and manage climate risk with devastating consequences for their operations, financial performance, and success in the marketplace. This chapter analyzes how climate impacts (physical risk) and changes in public policy and the market (transition risk) are reshaping the private sector; and how these impacts force a rethink of enterprise risk management at the company level.

Part I, Chapter 2: Creative destruction: shifting to a new climate economy

If the opportunity of shifting to a new climate economy is so staggering, why are we collectively failing to seize it and hence to accelerate this necessary transition? Given the systemic nature of the problem to be solved, it comes as no surprise that systemic barriers will stand in the way of solving it. This chapter explores critical such barriers and presents attempts by the private sector to overcome them. Starting with CSR all the way to CEO activism, this chapter will show that these attempts, while useful, are still embedded within the system that has led to – and sustains – the climate predicament we are currently in. This chapter concludes by making the case that to shift away from this system will require a portfolio of adapted solutions that are oriented to moving away and ultimately destroying this system in order to bring one about that is consistent with the new climate economy.

Part I, Chapter 3: The Greatest economic opportunity in history: building the new climate economy

Historically, climate action has been associated with cost and risk. Increasingly, studies and examples show that the transformation needed to avert the worst impacts of climate change are better understood as economic opportunities for companies. In all sectors of the economy, and especially in the high-emitting ones described in this chapter (energy, transport, and food and land-use), figures show companies and entire subsectors that have availed of these opportunities while predictions show that these upward trends will continue to strengthen, even in the context of a world affected by the global COVID-19 pandemic. Corporate goals – economic, strategic, environmental, and social – are all advanced by the opportunity of building a new climate economy, suggesting that companies that seize them will be at an advantage in this new paradigm.

Part II, Chapter 4: Building the low-carbon, resilient and inclusive economy

The new corporate climate leadership has three components. First, companies must decarbonize to avoid unmanageable climate change. This means reducing greenhouse gas emissions by 45% below 2010 levels by 2030 and achieving net-zero emissions by 2050.[31] Second, companies must build socio-ecological resilience to manage the unavoidable climate impacts we are already experiencing. This means investing in the six so-called capital assets – human, social, natural, physical, financial, and political – that are the key building blocks of resilience. Third, companies must create a shared prosperity by building a just transition from the high carbon and deeply unequal society of today to the new economy of tomorrow. This chapter presents the innovations that companies can use to decarbonize energy, transport, and food systems; guidance on how to apply the six capital assets inside individual companies, across complex global supply chains, and within frontline communities; and advice on how to secure a just transition.

Part II, Chapter 5: *Current corporate leadership: how commitments and inconsistencies shape the private sector response to climate crisis*

Over six thousand companies and investors, representing $36 trillion in revenue, have committed to climate action. Many have committed to absolute emissions reductions across their full value chain consistent with holding global mean temperature rises below 2°C above pre-industrial levels. These companies are pledging to procure 100% of their energy from renewable sources, draw the era of the internal combustion engine to a close, end commodity-driven deforestation, and change the way they invest and spend their money. However, this corporate climate leadership is not all that it may at first appear to be. Commitments are plagued by inconsistencies. While many companies have real plans to back up their pledges, others are being accused of setting rhetorical commitments, devoid of real investment, and lacking in implementation strategies. Corporate commitments are also undermined by a key contradiction, namely the problems that occur when an aspiration to decarbonize collides with the business instinct to pursue ever-increasing levels of growth. This chapter explores the strengths, inconsistencies, and contradictions of the current corporate climate leadership.

Part III, Chapter 6: *The innovation agenda: reimagining the climate compatible business*

The private sector has made great strides in recent years to understand and respond to the climate crisis. Companies around the world and across industrial sectors have committed to reducing greenhouse gas emissions, advocated for ambitious climate policies, and worked to resource supply chain decarbonization and resilience. However, current levels of ambition remain insufficient and inconsistent, and so corporate leadership needs to be dynamic. This chapter presents a framework for injecting new dynamism into corporate climate leadership through a three-point framework called *act, enable,* and *accountability*. Companies are encouraged to *act* to improve their diagnosis of climate risk, set more ambitious and robust decarbonization targets, and complement their emissions reductions pledges with new commitments on resilience. Companies are invited to *enable* greater climate ambition by addressing the inequality that amplifies climate risk, empowering communities' disproportionality impacted by climate change, and engaging in the process of debanking, reinvesting, and scaling finance to support the transition to a new climate economy. Companies are urged to embrace *accountability* by working to shape a policy enabling environment catalytic of business leadership on climate; conduct their business with integrity, ensuring all business practices are aligned with climate commitments; and scale-up disclosure to provide transparent information on risk and resilience to the market.

Part III, Chapter 7: Financial rewards: rethinking assets and liabilities in the context of climate change

Finance plays a unique role in unlocking the potential for climate leadership: it is an enabler and a transformer for companies in all sectors and for all the different kinds of actions we have outlined so far in this volume. First of all, finance in the context of climate represents a risk for companies insofar as they are exposed to sectors and activities that will become increasingly divested. Secondly, whether it is investing in innovation, developing new products or business models, building resilience in supply chains and communities, finance serves as the key to make these changes possible. Finally, organizations in the financial services sector can push transformation in all sectors by scaling up disclosure reporting requirements as well as aligning their own policies with climate-compatible standards.

Part III, Chapter 8: Corporate governance, purpose, and stakeholders: transforming companies to transform the economy

Calls for companies to act with purpose are becoming more prevalent, but what does this mean? And how can this contribute to establishing the new corporate climate leadership? If the purpose of a company is understood as the reason it exists, that is, the function it has in our society, it cannot be understood apart from its relationships with its stakeholders. Moreover, in order to ensure that these relationships are fair and non-exploitative, governance structures need to be implemented. This chapter explores how this conception of stakeholder relationships, buttressed by a strong vision of purposeful governance, can provide the tools to transform a company and orient it toward the new climate economy.

Part III, Chapter 9: From evolution to revolution: civil society and rethinking business in a new climate economy

Amongst the key stakeholders of companies seeking to lead the transformation of the economy is civil society, varied and heterogeneous as it is. By outlining a short history of civil society influence and mobilization on key issues – and not just climate – this chapter shows the critical role of these organizations and hence the opportunity for companies to engage with them. This chapter also explores the risks attached to strategy, specifically since reputation and trust are high stakes in an era of transparency.

References

1. Landrigan, P., Fuller, R., Acosta, N., Adeyi, O., Arnold, R., Basu, N., Baldé, A., Bertollini, R., Bose-O'Reilly, S., Boufford, J., Breysse, P., Chiles, T., Mahidol, C., Coll-Seck, A., Cropper, M., Fobil, J., Fuster, V., Greenstone, M., Haines, A. and Zhong, Ma. (2017). "The Lancet Commission on Pollution and Health". *The Lancet 391*, doi: 10.1016/S0140-6736(17)32345-0. https://www.thelancet.com/pdfs/journals/lancet/PIIS0140-6736(17)32345-0.pdf
2. UNDRR (United Nations Office for Disaster Risk Reduction) (2020) *The Human Cost of Disasters: An Overview of the Past 20 Years 2000–2019*. Geneva: UNDRR. https://reliefweb.int/report/world/human-cost-disasters-overview-last-20-years-2000-2019
3. WEF (World Economic Forum) (2020) *The Global Risks Report 2020*. Geneva: WEF. http://www3.weforum.org/docs/WEF_Global_Risk_Report_2020.pdf
4. Dietz, S., Bowen, A., Dixon, C. and Gradwell, P. 2016. "Climate Value at Risk of Global Financial Assets". *Nature Climate Change 6*: 676–679. http://eprints.lse.ac.uk/66226/1/Dietz_Climate%20Value%20at%20risk.pdf
5. TCFD (Task Force on Climate-related Financial Disclosures) (2016) *Phase I Report of the Task Force on Climate-Related Financial Disclosures*. London: TCFD. https://www.fsb-tcfd.org/wp-content/uploads/2017/06/FINAL-2017-TCFD-Report-11052018.pdf
6. Porter, K.A. and Yuan, J.Q., (eds). (2020) *A Roadmap to Resilience Incentivization*. Multi-Hazard Mitigation Council. Washington, DC: National Institute of Building Sciences. https://www.nibs.org/files/pdfs/NIBS_MMC_RoadmapResilience_082020.pdf
7. Levin, K. and Tirpak, D. (2018) *2018: A Year of Climate Extremes*. Washington, DC: World Resources Institute. https://www.wri.org/blog/2018/12/2018-year-climate-extremes
8. Bodnar, P. and Grbusic, T. (2020) "Your Climate Disaster Tax Bill Is Growing". In: *The New York Times*, June 23 2020. New York: The New York Times Company. https://www.nytimes.com/2020/06/23/opinion/climate-change-financial-disaster.html
9. Cameron, E. et al. (2019) *Companies and Climate Resilience: Mobilizing the Power of the Private Sector to Address Climate Risks*. Den Haag, Netherlands: Red Cross Red Crescent Climate Centre. https://static1.squarespace.com/static/59486a61e58c62bbb5b3b085/t/5d41ec40262f7500014c8ed3/1564601409741/RCCC±Private±sector±working±paper.pdf
10. Bradshaw, C.J.A., Ehrlich, P.R., Beattie, A., Ceballos, G., Crist, E., Diamond, J., Dirzo, R., Ehrlich, A.H., Harte, J., Harte, M.E., Pyke, G., Raven, P.H., Ripple, W.J., Saltré, F., Turnbull, C., Wackernagel, M. and Blumstein, D.T. (2021) "Underestimating the Challenges of Avoiding a Ghastly Future". *Frontiers in Conservation Science 1*:615419. https://www.frontiersin.org/articles/10.3389/fcosc.2020.615419/full
11. The White House. (1965) *Restoring the Quality of Our Environment: Report of the Environmental Pollution Panel President's Science Advisory Committee*. Washington, DC: US Government Printing Office. https://carnegiedge.s3.amazonaws.com/downloads/caldeira/PSAC,%201965,%20Restoring%20the%20Quality%20of%20Our%20Environment.pdf

12. Pacheco-Vega, R. and Parizeau, K. (2018) "Doubly Engaged Ethnography: Opportunities and Challenges When Working With Vulnerable Communities". *International Journal of Qualitative Methods* 17(1). December 1, 2018. Available from: https://doi.org/10.1177/1609406918790653
13. Coates, T. (2018) *We Were Eight Years in Power: An American Tragedy*. New York: One World Publishing.
14. Christensen, J. and Olhoff, A. (2019) *Lessons from a Decade of Emissions Gap Assessments*. Nairobi: United Nations Environment Programme. https://wedocs.unep.org/bitstream/handle/20.500.11822/30022/EGR10.pdf?sequence=1&isAllowed=y
15. Wei, D., Cameron, E., Harris, S., Prattico, E., Scheerder, G. and Zhou, J. (2016) *The Paris Agreement: What It Means for Business*. New York: We Mean Business. https://www.bsr.org/en/our-insights/report-view/the-paris-agreement-on-climate-what-it-means-for-business
16. Dauvergne. (2016) *Environmentalism of the Rich*. Cambridge, MA: The MIT Press.
17. WFP (World Food Programme) (2020), "Global Network against Food Crisis and Food Security Information Network (FSIN)". *Global Report on Food Crises 2020*. Rome: WFP. https://docs.wfp.org/api/documents/WFP-0000114546/download/?_ga=2.48756902.125074844.1587494438-119669374.1548357384
18. WFP (World Food Programme) (2020) "COVID-19 Will Almost Double Acute Hunger by End of 2020". In: *World Food Programme Insight*. Rome: WFP. https://insight.wfp.org/covid-19-will-almost-double-people-in-acute-hunger-by-end-of-2020-59df0c4a8072
19. FAO, IFAD, UNICEF, WFP and WHO (2019) "The State of Food Security and Nutrition in the World 2019". In: *Safeguarding against Economic Slowdowns and Downturns*. Rome: FAO. http://www.fao.org/3/ca5162en/ca5162en.pdf
20. IPCC (Intergovernmental Panel on Climate Change) (2019) *Climate Change and Land: An IPCC Special Report on Climate Change, Desertification, Land Degradation, Sustainable Land Management, Food Security, and Greenhouse Gas Fluxes in Terrestrial Ecosystems* [P.R. Shukla, J. Skea, E. Calvo Buendia, V. Masson-Delmotte, H.-O. Pörtner, D. C. Roberts, P. Zhai, R. Slade, S. Connors, R. van Diemen, M. Ferrat, E. Haughey, S. Luz, S. Neogi, M. Pathak, J. Petzold, J. Portugal Pereira, P. Vyas, E. Huntley, K. Kissick, M. Belkacemi, J. Malley, (eds.)]. Geneva: IPCC. https://www.ipcc.ch/srccl-report-download-page/
21. Development Initiatives (2018) *2018 Global Nutrition Report: Shining a Light to Spur Action on Nutrition*. Bristol, UK: Development Initiatives. https://www.unicef.org/press-releases/2018-global-nutrition-report-reveals-malnutrition-unacceptably-high-and-affects
22. UNICEF (2018) *Levels & Trends in Child Mortality 2018 Estimates Developed by the UN Inter-Agency Group for Child Mortality Estimation*. New York: United Nations Inter-agency Group for Child Mortality Estimation. https://weshare.unicef.org/Package/2AMZIFV567G1#/SearchResult&ALID=2AMZIFV567G1&VBID=2AM4WR6L6YZG&POPUPPN=1&POPUPIID=2AMZIFVG3LVA
23. Lancet Commissions (2018) *The Lancet Commission on pollution and health. Lancet 2018*; 391: 462–512. https://www.thelancet.com/journals/lancet/article/PIIS0140-6736(18)31039-0/fulltext
24. The Economist (2018) "The Black Hole of Coal: India Shows How Hard It Is to Move beyond Fossil Fuel". *The Economist 432*: 9162.
25. McNeil, D. (2018) Ticks and Mosquito Infections Spreading Rapidly, CDC Find. In: *The New York Times*, New York: The New York Times. Published on May 1, 2018.
26. International Federation of Red Cross and Red Crescent Societies (2017) *Framework for Climate Action Towards 2020*. Geneva: IFRC. http://www.climatecentre.org/downloads/files/CLIMATE%20FRAMEWORK%20FULL.pdf

27. WMO (World Meteorological Organization) (2019) *WMO Statement on the State of the Global Climate in 2018*. Geneva: WMO. https://library.wmo.int/index.php?lvl=notice_display&id=20799#.XqjFYy-z3OQ
28. Pierre-Louis, K. (2018) "These Billion-Dollar Natural Disasters Set a US Record in 2017". In: *The New York Times*, New York: The New York Times. Published on January 8, 2018.
29. Editorial Board. (2018) *We Won't Stop California's Wildfires if We Don't Talk about Climate Change*. The Washington Post. 8th August 2018.
30. Caesar, L., McCarthy, G.D., Thornalley, D.J.R., Cahill, N. and Rahmstorf, S. (2021) "Current Atlantic Meridional Overturning Circulation weakest in Last Millennium". *Nature Geoscience, doi*: 10.1038/s41561-021-00699-z. Published, 25 February, 2021.
31. IPCC (Intergovernmental Panel on Climate Change) (2018) "Summary for Policymakers". In: *Global Warming of 1.5°C. An IPCC Special Report on the Impacts of Global Warming of 1.5°C above Pre-industrial Levels and Related Global Greenhouse Gas Emission Pathways, in the Context of Strengthening the Global Response to the Threat of Climate Change, Sustainable Development, and Efforts to Eradicate Poverty*. Geneva: IPCC. https://www.ipcc.ch/site/assets/uploads/sites/2/2019/06/SR15_Summary_Volume_Low_Res.pdf
32. IPCC (Intergovernmental Panel on Climate Change) (2018) *Global Warming of 1.5°C. An IPCC Special Report on the Impacts of Global Warming of 1.5°C above Pre-industrial Levels and Related Global Greenhouse Gas Emission Pathways, in the Context of Strengthening the Global Response to the Threat of Climate Change, Sustainable Development, and Efforts to Eradicate Poverty*. Geneva: IPCC. https://www.ipcc.ch/site/assets/uploads/sites/2/2019/06/SR15_Full_Report_Low_Res.pdf
33. UNEP (United Nations Environment Programme) (2018) *The Emissions Gap Report 2018*. Nairobi: UNEP. https://www.unenvironment.org/resources/emissions-gap-report-2018
34. Hawken, P. (ed). (2017) *Drawdown: The Most Comprehensive Plan Ever Proposed to Reverse Global Warming*. London: Penguin Books.
35. Mulligan, J., Ellison, G., Levin, K., Lebling, K. and Rudee, A. (2020) 6 Ways to Remove Carbon Pollution from the Sky. Washington, DC: The World Resources Institute. https://www.wri.org/insights/6-ways-remove-carbon-pollution-sky
36. Cameron, E. and Nestor, P. (2018) *Climate and Human Rights: The Business Case for Action*. San Francisco: BSR. https://www.bsr.org/reports/BSR_Climate_Nexus_Human_Rights.pdf
37. Cameron, E. (2019) "Business Adaptation to Climate Change and Global Supply Chains". *Background Paper for the Global Commission on Adaptation*. The Hague: Global Commission on Adaptation. https://gca.org/reports/business-adaptation-to-climate-change-and-global-supply-chains/
38. International Labour Organization (ILO). (2016) *A Just Transition to Climate-Resilient Economies and Societies: Issues and Perspectives for the World of Work*. Geneva: ILO. https://www.ilo.org/wcmsp5/groups/public/—ed_emp/—gjp/documents/publication/wcms_536552.pdf
39. United Nations. (1948) *Universal Declaration of Human Rights*. Paris: United Nations. https://www.un.org/en/about-us/universal-declaration-of-human-rights
40. United Nations General Assembly. (1966) "International Covenant on Economic, Social, and Cultural Rights". *Treaty Series 999(1966)*: 171. https://www.ohchr.org/Documents/ProfessionalInterest/cescr.pdf
41. Office of the United Nations High Commissioner for Human Rights. (2008) *The Right to Health: Fact Sheet Number 31*. Geneva: OHCHR. https://www.ohchr.org/Documents/Publications/Factsheet31.pdf

42. Thompson, M. and Rayner, S. (1998) "Cultural Discourses". In: *Human Choice and Climate Change: The Societal Framework*, edited by [S. Rayner and E. L. Malone (eds.)], Vol. 1, pp. 265–343. Columbus, OH: Battelle Press.
43. UK Government, Department of Business, Energy, and Industrial Strategy; Press Release, 30 March 2021, https://www.gov.uk/government/news/third-of-uks-biggest-companies-commit-to-net-zero#:~:text=UK%20FTSE100%20companies%20who%20have,become%20net%20zero%20by%202040.
44. Hsu, A., Widerberg, O., Weinfurter, A., Chan, S., Roelfsema, M., Lütkehermöller, K. and Bakhtiari, F. (2018) "Bridging the Emissions Gap – The Role of Non-state and Subnational Actors". In: *The Emissions Gap Report 2018. A UN Environment Synthesis Report*. Nairobi: United Nations Environment Programme. https://wedocs.unep.org/bitstream/handle/20.500.11822/26093/NonState_Emissions_Gap.pdf?sequence=1
45. A sample of commitments taken by 1600 companies can be accessed through "We Mean Business". https://www.wemeanbusinesscoalition.org
46. United Nations Framework Convention on Climate Change; Article: February 8th 2021. https://unfccc.int/news/un-secretary-general-calls-for-exponential-growth-in-global-coalition-to-achieve-net-zero-emissions

PART I
Climate risks, response, and rewards for the 21st-century company

1
CLIMATE CHANGE: THE HIGHEST IMPACT RISK FOR GLOBAL BUSINESS

The Financial Stability Board (FSB) represents ministries of finance and central banks from across the G20, including the Federal Reserve and the European Commission. The FSB has identified climate change as a significant risk to the economy, defining risk in two broad categories – physical and transition. The total cost of the climate crisis to the global economy could be as high as US$24 trillion by 2030,[1] and as much as US$43 trillion between now and the end of the century.[2] The risks are global, affecting every geography and sector of the economy; and broad, with impacts disrupting every phase of the value chain from extraction of primary resources to consumption and even disposal of goods and services. This chapter analyzes how climate impacts (physical risk) and changes in public policy and the market (transition risk) are reshaping the private sector; and how these impacts force a rethink of enterprise risk management at the company level. The chapter offers a vital innovation – a three-dimensional understanding of climate risk – as the key to unlocking corporate resilience in the face of the climate crisis.

Physical risks from climate change

Climate risk is determined by the existence of physical hazards, exposure to those hazards, and underlying vulnerability. Climate hazards refer to the possible future occurrence of natural or human-induced physical events that may have adverse effects on vulnerable and exposed elements. Exposure refers to an area in which hazard events may occur. In other words, a hurricane moving slowly through the mid-Atlantic may be influenced by climate change, but it does little physical damage to human systems unless it makes landfall and passes through population centers. It is the presence of people; livelihoods; environmental services and resources; infrastructure; or economic, social, or cultural assets that determine whether a hazard may become a risk. Vulnerability is the propensity of exposed elements, whether people, ecosystems, biodiversity, economic sectors, complex supply chains or individual companies, to suffer adverse effects when climate-related hazards occur. It is the underlying weaknesses in their own systems that exacerbate risk.[3]

The Intergovernmental Panel on Climate Change (IPCC) has been tasked by the United Nations with providing objective, scientific information relevant to understanding the global climate system. It produces periodic assessments of the full spectrum of published, peer-reviewed science to determine the state of knowledge on climate change. The most recent IPCC assessment concluded that human activities are estimated to have already caused approximately 1.0°C of global warming above pre-industrial levels, with likely warming of 1.5°C by 2050. The consequences of this warming are already evident in the shifting patterns of climate hazards, including:

- An increase in intensity and frequency of extreme weather events, including hurricanes, wildfires and heatwaves. For example, 50–80% of the globe is projected to experience significantly more intense, longer and more frequent hot extremes than historically recorded.
- Alterations to land ecosystems and some of the services they provide. This includes fragmentation and destruction of habitat leading to dramatic species loss and extinction. Forest fires and the spread of invasive species will also contribute to biodiversity loss. Shifts in climate zones including the expansion of arid and contraction of polar zones are projected under high-emission scenarios.
- Alterations to ocean ecosystems and the species they support due to increases in ocean temperature, elevated ocean acidity, and decreases in ocean oxygen levels.
- An increase in heavy precipitation in regions susceptible to flooding and an increase in drought in historically dry regions. As a result of global warming, the number of record-breaking rainfall events globally has increased significantly by 12% during the period 1981 to 2010 compared to those expected due to natural multi-decadal climate variability. A recent analysis of 4,500 droughts found increased drought frequency over numerous parts of the globe, including engines of the world economy such as the Eastern United States and north-eastern China.
- Sea level rise of up to 0.77 m by 2,100 with significant consequences for coastal areas.[4]

These are not just projections for the future but are increasingly a real-time reflection of shifting socio-ecological systems. Natural disasters doubled during 2000–2019 when compared to 1980–1999. Climate-related hazards account for the difference, rising from 3,600 to 6,700 during this 20-year period. The impacts on human systems are substantial. In addition to causing financial losses in excess of US$3 trillion, hazards lead to significant human systems impacts. Climate change is already elevating health risks such as vector and water-borne disease and heat-related morbidity; exacerbating hunger and food insecurity; driving higher rates of water stress; undermining poverty alleviation; and affecting livelihoods, particularly in climate-sensitive sectors such as agriculture.[5]

The stability of our food system is projected to decrease as global mean temperatures rise and changing precipitation patterns alter the start and end of growing seasons; contribute to crop yield reductions, impact livestock productivity; reduce freshwater availability; and increase the incidence of pests and disease. Increased atmospheric CO_2 levels can also lower the nutritional quality of crops. Drylands are expected to experience reductions in crop and livestock productivity, with continued rising temperatures expected to impact global wheat yields by about 4–6% reductions for every degree of temperature rise. A recent study shows that between 18% and 43% of the explained yield variance of four crops (maize, soybeans, rice, and spring wheat) is attributable to extremes of temperature and rainfall. Changes in our food system will exacerbate vulnerability in human systems. Crop and economic models project up to 29% increases in cereal prices in 2050 due to climate change. Increases may expose up to 183 million additional people to hunger, lead to increased displacement, disrupted food chains, threatened livelihoods, and contribute to conflict.[6]

Between 2000 and 2019, there were 510,837 deaths and 3.9 billion people affected by 6,681 climate-related disasters. This compares with 3,656 climate-related events in the period 1980–1999.[7] Asia suffered the highest number of disaster events during this period – over 3,000 – and eight of the top ten countries suffered by disaster events are in Asia. Floods have the highest impacts in Asia, with 41% of all flooding events and a total of 1.5 billion people affected, accounting for 93% of people affected by floods worldwide. China experienced a wide variety of over 500 disaster events during this period, with flooding affecting a total of 900 million Chinese people over the two decades.[7]

Rapid development over the past 40 years means that South and East Asia are heavily exposed to hazards because of their large coastal populations in low-lying areas; populations that form the customer and employee base of the world; and many of the workers in global supply chains. A report prepared by the New Economics Foundation predicts that the "human drama of climate change will largely be played out in Asia, where over 60 percent of the world's population lives, over half of those live near the coast, making them directly vulnerable to sea-level rise".[8] Considering that in 2014, East Asia accounted for 60% of the container volume among the 100 largest ports in the world – equivalent to four times the volume of European ports and six times US ports – the implications of climate-related hazards in the region for global value chains cannot be underestimated.[9]

In the United States, disaster losses from wind, floods, earthquakes, and fires now average $100 billion per year, and in 2017, exceeded $300 billion – 25% of the $1.3 trillion building value put in place that year.[10] Climate-related disaster losses are growing at a rate of 6% a year and costing the US economy an average of $100 billion annually.[11] The 2017 Atlantic hurricane season was the costliest on record causing over $290 billion damages in Harvey and Irma alone, while in 2018, Hurricanes Florence and Michael were just two of 14 "billion-dollar disasters" in the United States.[12] The Mendocino Complex Fire of 2018 spread to more than 300,000 acres, becoming the largest fire ever recorded in California. Over the same period, more than US$3 billion of damage was inflicted across five mid-western US states due to flooding. In Nebraska alone, the flooding caused more than US$1 billion in damages, with more than 2,000 homes and 340 businesses lost.[12] And in the year through May 2019, the United States experienced its wettest 12 months on record, including devastating floods affecting 14 million people in the Midwest and South.[13] This is just the beginning. BlackRock calculates a 275% increase in major hurricane risk by 2050, raising the stakes still further.[14] This has led the US Commodity Futures Trading Commission to conclude that climate change is already impacting nearly every facet of the economy, including infrastructure, agriculture, residential and commercial property, as well as human health and labor productivity. The Commission has warned that climate change could undermine the economy's ability to generate employment and income, cause disorderly price adjustments in various asset classes, and ultimately disrupt proper functioning of financial markets.[13]

Other engines of the global economy are also severely impacted. The European Union is the second-largest economy in the world in nominal terms, after the United States, and the third one in purchasing power parity terms, after China and the United States. The deadliest climate events in 2019 were the summer heatwaves that affected Europe with over 2,500 deaths.[15] Meanwhile, analysis of the wildfire damages and losses in Australia in 2019 and 2020 expect the final cost to exceed US$100 billion.[16]

Transition risks

The shift to a low-carbon economy, designed to dramatically reduce greenhouse gas emissions consistent with holding global mean temperature rise below 1.5°C above pre-industrial levels is not without its own range of risks. These "transition risks" include policy, legal, technological, and broader market drivers that build a new economy and disrupt business as usual. This could include the introduction of a price on carbon to the economic obsolescence of entire asset classes because of changing consumer preferences (i.e. stocks in fossil fuel companies and the value of high-carbon commodities such as crude oil). Transition risks may lead to economic losses for some sectors and stranded capital for the financial sector.[17]

The policy enabling environment is changing rapidly as a consequence of the landmark Paris Agreement in December 2015. Almost 200 countries signed the Agreement, and 189 of those now have national climate action plans, with economy-wide greenhouse gas emissions reductions targets covering a range of sectors, including energy, transportation, land use, and the built environment.[18]

The European Union Green Deal is the most ambitious of these, with the goal of making the EU climate neutral in 2050 and an interim greenhouse gas (GHG) reductions target of 50%–55% by 2030. In the United States, the Biden Administration has committed to economy-wide net-zero emissions by 2050, with interim goals of net-zero emissions in all new buildings by 2030, and in the electricity sector by 2035.[19] Thirty-two US states have climate action plans, including California, with a target of net-zero emissions by 2045, and 40% GHG reductions by 2030.[20] China has pledged to reach carbon neutrality by 2060, a massive geopolitical shift as well as a significant economic undertaking as China has long argued that its status as a developing economy shielded it from the emissions reductions obligations of industrialized nations.[21] The United Kingdom, the sixth-largest economy in the world and the host of a key climate summit in November 2021, has issued a 10-point plan for a "green industrial revolution" and pledged to reduce greenhouse gas emissions at least 68% by 2030, compared to 1990 levels and achieve carbon neutrality by 2050.[22] More than 60 other countries have pledged carbon neutrality by 2050, a significant catalyst for broader economic transition.[23]

Broader market shifts are also driving the transition, notably the dramatic increase in private sector climate commitments since 2014. Today, over 6,000 companies and investors, representing US$36 trillion in revenue, have committed to climate action. These companies, from 120 countries, make up half the global economy.[24] Financial institutions with cumulative assets of at least $47 trillion under management (representing 25% of the global financial market) have set climate-related targets for their portfolios.[25]

These commitments, from both the public and private sectors, have consequences for high-carbon industries. The United States consumes 40% less coal than at its peak in 2005. Ten years ago, coal produced 50% of the nation's electricity; today, it accounts for 25%. Coal consumption is projected to fall by another 25% in the coming decade. As a result, there are 239 coal-fired plants in the United States, down from 600 in 2007.[26] In all net-zero pathways, most of the nearly 700 mines will close by 2030, impacting all coal-producing regions.[27]

Together the top four coal mining companies in the United States have lost over $44 billion in market cap since 2011 – a 99% decrease in valuation – as a result of changing policy environments and their corporate consumers moving away from coal-fired energy.[28] US coal stocks lost over half their value in 2019 alone,[29] and coal capacity under development is down 62% globally since 2015.[30] In the United Kingdom, coal has fallen from providing 40% of UK energy needs in 2012 to providing just 5% in 2018, a decline of 88% over just six years. In that same time period, generation for renewables came online to serve this gap.[31] This has led some analysts to conclude that market forces have already reduced coal consumption in industrialized countries past a point of no return.[32] Coal companies are not the only corporations at risk due to this transition to renewable energy.

Other fossil fuel industries are at the beginning of the transition as well. The value of oil is declining just as oil companies grapple with unprecedented debt and threats. The world's largest listed oil companies lost almost $90 billion in value during the first nine months of 2020. In September 2020, 255 oil rigs were operating in the United States, down from the 2011 high of 2,026 operating rigs. This led to 118,000 energy workers losing their jobs in the United States between March and July 2020, a 15.5% decline in the workforce.[33]

Oil is worth $16 trillion of capital and employs over 10 million people. Twenty-six countries depend on oil income for 5% or more of their GDP; the average among the 26 is 18%. Saudi Aramco, the largest oil company in the World, had a net income of $111 billion in 2018, twice as much as Apple. Governments are now beginning to revisit the support they provide to this industry, support that is increasingly in direct conflict with national climate commitments. The transition risks are therefore significant for countries as well as corporations.[34] Governments currently offer US$427 billion per year in pre-tax subsidies to fossil fuels; and indirect subsidies by building infrastructure, managing environmental degradation and offering healthcare to treat air pollution illnesses.[35] What happens to the profitability of these companies when these subsidies begin to drain away?

This unprecedented transition in energy systems away from fossil fuels and toward renewable energy has significant implications for the financial services sector too, who are now finding their assets turning into liabilities. Thirty-five banks from Canada, China, Europe, Japan, and the United States have funneled USD $2.7 trillion into fossil fuels since 2016. Finance to 100 of the biggest expanders of coal, oil, and gas fell by 20% between 2016 and 2018, but last year bounced back 40%. Twenty-six of these companies now have policies restricting coal finance, and 16 have policies to restrict finance to some oil and gas sectors.[36] The speed at which these new policies take effect will determine the scale and speed of transition risks for both the financial services industry and the high-carbon sectors they are currently invested in. One estimate for stranded capital from fossil fuel assets suggests a potential global loss of wealth between $1 trillion and $4 trillion.[37] In an alternative estimate, current stranded assets within fossil fuel companies range between $250 billion and $1.2 trillion – depending on how fossil fuel firms respond to global emissions reductions.[38]

Corporate climate risk

The World Economic Forum (WEF) has consistently ranked climate change as the highest material risks to global business in both likelihood and impact.[39] Complex supply chains experience climate risk, from those cultivating natural resources or extracting raw materials to those producing manufactured goods, transporting products, or ultimately selling finished articles to consumers. Climate risk cannot be easily confined to one particular segment of a company. The authors have examined 64 different business risk vectors across four strategic categories – operations and supply chain; human resources; compliance and legal; and finance – and have found both physical and transition risk to be threat multipliers in each.

Operational and supply chain risk

Leading companies understand that success is not based on the strength of operations within their own four walls but is really determined by the depth and reliability of their supply chain. The real competition is not company against company, but rather supply chain against supply chain.[40] Supply chains create value by being reliable and responsive in matching demand and supply. Reliability is delivering the right product in the right quantity at the right time to the right place at the lowest cost. Responsiveness is the ability to respond quickly to changing market conditions.[41] Modern supply chains span the globe and involve many suppliers, contract manufacturers, distributors, logistics providers, original equipment manufacturers, wholesalers and retailers. The supply chain includes the inbound elements, featuring all the processes and suppliers responsible for furnishing the company with materials and parts; the internal processes within companies that convert inputs into manufactured goods and services; and the outbound or customer-facing elements that focus on distribution processes and customers.[42] This vast network consists of materials and flows and depends on the locations of suppliers, sub-suppliers and service providers. A car may contain up to 50,000 unique parts, manipulated in one form or another in dozens of countries and crossing borders many times over.[43] As much as 80% of global trade is embedded in global supply chains, including trade in intermediate goods and services of about US$12 trillion.[44]

Climate change can impact operations and supply chains in a number of ways, including:

- Location risk: Some population centers are becoming increasingly susceptible to intense and frequent climate events. This may lead companies to question to the wisdom of placing production, distribution or retail outlets in these locations.
- Infrastructure and logistics risks: Climate change is a growing threat to infrastructures such as roads, bridges, ports and public utilities such as those dealing with electricity generation, water and waste treatment. Damage to infrastructure has an impact on the owning company, but the disruption to the wider economy from infrastructure outages is even greater.
- Production and procurement shortfall risk: An increasingly volatile climate may prevent a company from procuring the raw materials it needs to produce goods and services and may halt production and manufacturing if factories are affected.

Human resources risk

Companies tend to think about human resources risks through the lens of workplace safety and talent management. The former concerns the threat of accidents or how broader health impacts for employees (i.e. through exposure to hazardous conditions or compounds); whereas the latter relates to a company's ability to recruit and retain staff. Increasingly, climate change is forcing companies to understand their workforce and how certain groups of people face asymmetrical impacts from climate risk.

Compliance and legal risk

Compliance and legal issues risks arise when companies fail to adhere to laws and regulations designed to enhance adaptive capacity; from liability arising from climate-related lawsuits; and from failure to fully disclose climate risks through mandatory reporting mechanisms. Climate activists are increasingly turning to the courts to drive the climate ambition they see as being absent from boardrooms. In recent years, a variety of laws including those dealing with environmental damage, human rights violations, and breach of fiduciary or transparency, have been used in an attempt to bring polluters to account for climate change. More than 1,100 climate-related lawsuits have been filed in the United States,[45] at a rate of approximately 20 new lawsuits each year.[46] In the past 15 years, 64 such cases have been brought in countries other than America.

The most high-profile litigation has centered on complaints advanced by state attorneys general for violations of state securities laws, among other allegations, against a fossil fuel legacy firm for its alleged failure to adequately disclose material climate-related risks to investors. In securities law, future legal risks likely will involve decisions about whether climate-related risk factors are material enough to require disclosure, as well as the adequacy of disclosures.[47] Shareholder activism is increasingly focused on whether officers, directors and other fiduciaries may be violating fiduciary duties by investing in, or failing to disinvest in, various carbon-intensive or otherwise highly exposed assets, companies, and industries.[48] Compliance and legal issues also arise from potential liability for failing to adequately disclose material physical risks on debt offerings and other contracts.[49]

Financial risk

The key concern for companies is that climate change will turn assets into liabilities; increase supply chain disruption, thus undermining competitive advantage; and ultimately lead to financial losses. Financial risk can be exacerbated by climate change in a number of ways:

- Company profits could fall as vulnerability and exposure to climate risk becomes known – possibly through an increased focus on the content of disclosure reports or resulting from unfavorable media coverage of a supply chain disruption. A study of 861 public companies found that with the announcement of a supply chain malfunction, such as production or shipment delays, the company's stock price tumbled nearly 9% on average. Furthermore, that stock lost 20% of its value within six months of its announcement.[42] Research conducted on a sample of 519 supply chain disruptions during the period 1989–2000 concluded that such disruptions are associated with an abnormal decrease in shareholder value of 10.28%.[41] In 2015, Henri de Castries, CEO of Axa, announced that the world's path of more than 4°C of global warming would create an uninsurable world of climate-related physical hazards. The lack of availability of capital, including from insurance after an extreme weather event, flood or sea-level rise could cause financial strain on businesses.[50]

- Profitability could also be hit due to diminished capital availability and higher credit risk, as investors and lenders refuse to make capital available to companies seen to be exposed and vulnerable to climate impacts or not managing them adequately. The Global Investor Coalition on Climate Change, which represents US$24 trillion in assets, expects companies to have climate change strategies in place. The California Public Employees' Retirement System, or CalPERS, for instance, requires from corporations it invests in that board members have climate expertise.[51] BlackRock, the largest asset manager in the world – with its US$5.1 trillion under management, equivalent to 4.3% of the world's gross domestic product (GDP) – announced that it would expect companies to provide assessments of how climate change would affect their business.[52]

- Large accounts may be at risk of significant procurers of goods and services begin to focus their purchasing on climate-friendly suppliers.[53] This, in part, explains the substantial decline in the coal industry in recent years as companies procure renewable energy in support of publicly stated climate commitments.

- Risks related to asset and commodity prices are particularly strong because of the impact of climate change on ecosystems services, food production and real estate. High variability in the price of raw materials driven by vulnerability and exposure to climate hazards may dramatically change inputs vital to production.[2]

- Reputational damage may result from a perceived failure to account for climate risk with implications for revenue, as customers and suppliers become hesitant to associate with a discredited brand. According to research by Morgan Stanley, millennials purchased from a sustainable brand twice as often as the total individual investor population. Millennials were three times more likely to have sought employment with a sustainably minded company and invested in companies targeting social/environmental goals twice as much as the total individual investor population. This is important as millennials are expected to make up 75% of the American workforce by 2025, and 90% of them have expressed an interest in pursuing sustainable investments as part of their retirement savings.[54]

Profiles in corporate climate risk

These risks are no longer merely projected in climate models. In recent years, they have become all too familiar in the experiences of some of the world's leading companies.

In 2011, the worst floods in more than 50 years struck Thailand, causing millions to become displaced or homeless, putting some 650,000 people at least temporarily out of work, and bringing activity at businesses, schools and hospitals to a standstill. Seventy-seven of Thailand's 84 provinces were affected, resulting in economic losses of more than US$45.7 billion or 13% of that year's GDP.[55] The floods had systemic impacts on the technology and automotive sectors. The industrial parks in central Thailand had become a cluster for making hard disks and their components. At the time, Thailand provided 45% of the worldwide hard drive production. When the floods inundated 877 factories, halting 30% of the global hard disk manufacturing, the personal computer industry faced a 35% shortfall in disk supplies in the fourth quarter of 2011.[42] Hewlett-Packard Technology, at the time the world's leading PC manufacturer, suffered in excess of US$4 billion in lost revenue as a consequence of the floods.[42] Thailand had also become a global center for producing car parts. Japanese carmakers t like Toyota and Honda were forced to cut back production at plants as far away as North America and suffered substantial losses because their Thai suppliers were underwater.[56]

Before October 2012, Verizon's telephone infrastructure depended on copper-based systems to support its landlines nationally, including in New York and New Jersey. But when Hurricane Sandy sent a surge of saltwater sweeping through the company's facilities, the 90,000 cubic foot cable vault suffered a catastrophic failure and the copper wiring dissolved. As a result, thousands of Verizon customers lost service; the company suffered reputational and operational damage; and, ultimately, the company lost approximately US$1 billion.[57] Verizon suffered these consequences because although its modeling recognized the likelihood of increased intensity and frequency of extreme weather events due to climate change and understood the company's exposure to these extreme weather events, it failed to address its underlying weaknesses or vulnerability, namely its dependence on physical capital and infrastructure that were incompatible with a changing climate.

In 2017, Hurricane Maria made landfall in Puerto Rico, leveling homes, flooding vast swaths of the island. Pharmaceuticals and medical devices constitute Puerto Rico's leading exports, and drug companies and device makers represent a US$15 billion stake there. Baxter International manufactures intravenous (IV) bags on the island – in fact, the Fortune 500 healthcare company constitutes more than 40% of the United States' IV solution market. When Hurricane Maria forced the shutdown of Puerto Rican plants, hospitals that relied on these products were unable to resupply. When an unusually severe flu season swept across the United States, hospitals were left scrambling for IV bags to care for dehydrated flu patients. In some cases, clinics nowhere near the storm suddenly found themselves paying up to a 600% markup.[58]

In 2019, PG&E announced the biggest utility bankruptcy in US history. A company with $71 billion in assets and $51 billion in debt, they were confronted with $30 billion in estimated wildfire liabilities due to their failure to maintain infrastructure and adapt their business in the face of increased intensity and frequency of wildfires in Northern California in 2017 and 2018.[59] They were ultimately forced into an $11 billion settlement and their share price collapsed by 90%.[60]

The transition is also taking a toll on established companies. In the United States, the top four mining companies – Peabody, Arch Coal, Alpha Natural and Walter Energy – have collectively lost over US$44 billion in market capitalization from their peaks in 2011 due to the availability of cheaper energy sources, changing consumer preferences, and a shifting regulatory environment.[61]

This has implications for the balance sheets of some of the largest financial institutions in the world. The 2011 floods in Thailand cost Swiss Re more than US$600 million in claims, but the total insured market loss could have been in the range of US$8 billion to US$11 billion.[62] JPMorgan Chase has provided US$269 billion in financing to the fossil fuel industry between 2016 and 2019. Bank of America, meanwhile, is the eighth biggest funder of coal power in the same period, with almost US$160 billion of financing, making it the largest non-Chinese coal power funder in 2019.[36]

Conclusion

Climate change is elevating physical risks, revealing transition risks, and fundamentally altering understanding of enterprise risk management inside individual companies and across the increasingly complex and fragile global supply chains upon which they depend.

The companies profiled in this chapter span the economy, drawn from sectors as diverse as food and agriculture, telecommunications, auotmotives, information and communications technology, pharmaceuticals, energy, and financial services. And yet, the commonality is clear. Their failure to properly understand climate change has amplified their business risk. They have little insight into the far reaches of their supply chains, the transportation system that connects them, the raw materials used, and the intersecting inequalities faced by the people who populate their supply chains. The creation of "just-in-time" supply chains means there are often few redundancies or contingencies to deal with disruption. The scope of the supply chain across so many geographies means there are multiple dependencies that elevate the risk of broader contagion when one part of the chain fails. And the global nature of climate change, with many events occurring simultaneously, means there are multiple exposures for companies to deal with.

In essence, they have failed to properly diagnose climate risk and are paying the price for that failure. Moving away from an outdated two-dimensional understanding of risk (focused on determining and seeking ways to reduce exposure to a hazard) toward a three-dimensional diagnosis that foregrounds vulnerability as a risk amplifier is vital to securing competitive supply chains in a global economy. Properly diagnosing climate risk using three dimensions is explored in greater detail in Chapter 6.

References

1. Dietz, S., Bowen, A., Dixon, C., and Gradwell, P. (2016) "Climate Value at Risk of Global Financial Assets". *Nature Climate Change 6*: 676–679. http://eprints.lse.ac.uk/66226/1/Dietz_Climate%20Value%20at%20risk.pdf
2. TCFD (Task Force on Climate-Related Financial Disclosures) (2016) *Phase I Report of the Task Force on Climate-Related Financial Disclosures.* London: TCFD. https://www.fsb-tcfd.org/wp-content/uploads/2017/06/FINAL-2017-TCFD-Report-11052018.pdf
3. IPCC (Intergovernmental Panel on Climate Change) (2012) "Summary for Policymakers". In: *Managing the Risks of Extreme Events and Disasters to Advance Climate Change Adaptation: A Special Report of Working Groups I and II of the Intergovernmental Panel on Climate Change.* Geneva: IPCC. https://www.ipcc.ch/report/managing-the-risks-of-extreme-events-and-disasters-to-advance-climate-change-adaptation/
4. IPCC (2018) "Summary for Policymakers". In: *Global Warming of 1.5°C. An IPCC Special Report on the Impacts of Global Warming of 1.5°C above Pre-Industrial Levels and Related Global Greenhouse Gas Emission Pathways, in the Context of Strengthening the Global Response to the Threat of Climate Change, Sustainable Development, and Efforts to Eradicate Poverty* [Masson-Delmotte, V., P. Zhai, H.-O. Pörtner, D. Roberts, J. Skea, P.R. Shukla, A. Pirani, W. Moufouma-Okia, C. Péan, R. Pidcock, S. Connors, J.B.R. Matthews, Y. Chen, X. Zhou, M.I. Gomis, E. Lonnoy, T. Maycock, M. Tignor, and T. Waterfield (eds.)]. In Press. Available from: https://www.ipcc.ch/sr15/download/
5. UNDRR (United Nations Office for Disaster Risk Reduction) (2020) *The Human Cost of Disasters: An Overview of the Past 20 Years 2000-2019.* Geneva: UNDRR. Available from: https://reliefweb.int/report/world/human-cost-disasters-overview-last-20-years-2000-2019
6. IPCC (2019) "Summary for Policymakers". In: *Climate Change and Land: An IPCC Special Report on Climate Change, Desertification, Land Degradation, Sustainable Land Management, Food Security, and Greenhouse Gas Fluxes in Terrestrial Ecosystems.* Geneva: IPCC. Available from: https://www.ipcc.ch/site/assets/uploads/2019/08/Edited-SPM_Approved_Microsite_FINAL.pdf
7. UNDRR (United Nations Office for Disaster Risk Reduction) (2020) *The Human Cost of Disasters: An Overview of the Last 20 Years (2000-2019).* Geneva: UNDRR. Available from: https://www.undrr.org/news/drrday-un-report-charts-huge-rise-climate-disasters
8. NEF (New Economics Foundation) (2007) *Up in Smoke: Asia and the Pacific – the Threat from Climate Change to Human Development and the Environment. Third Working Group on Climate Change and Development.* London: NEF. https://pubs.iied.org/pdfs/10020IIED.pdf
9. AAPA (American Association of Port Authorities) (2014). *World Port Rankings 2014.* Alexandria, VA: AAPA. https://www.aapa-ports.org/unifying/content.aspx?ItemNumber=21048
10. Porter, K., Dash, N., Huyck, C., Santos, J., Scawthorn, C., Eguchi, M., Eguchi, R., Ghosh, S., Isteita, M., Mickey, K., Rashed, T., Reeder, A., Schneider, P., and Yuan, J. (2019) *Natural Hazard Mitigation Saves: 2019 Report.* Washington, DC: Multi-Hazard

Mitigation Council/National Institute of Building Sciences. Available from: https://cdn.ymaws.com/www.nibs.org/resource/resmgr/reports/mitigation_saves_2019/mitigationsaves2019report.pdf

11. Porter, K.A. and Yuan, J.Q., (eds.) (2020) *A Roadmap to Resilience Incentivization. Multi-Hazard Mitigation Council*. Washington, DC: National Institute of Building Sciences. Available from: https://www.nibs.org/files/pdfs/NIBS_MMC_RoadmapResilience_082020.pdf
12. Levin, K. and Tirpak, D. (2018) *2018: A Year of Climate Extremes*. Washington, DC: World Resources Institute. Available from: https://www.wri.org/blog/2018/12/2018-year-climate-extremes
13. Climate-Related Market Risk Subcommittee (2020) *Managing Climate Risk in the U.S. Financial System*. Washington, DC: Commodity Futures Trading Commission, Market Risk Advisory Committee. Available from: https://www.cftc.gov/sites/default/files/2020-09/9-9-20%20Report%20of%20the%20Subcommittee%20on%20Climate-Related%20Market%20Risk%20-%20Managing%20Climate%20Risk%20in%20the%20U.S.%20Financial%20System%20for%20posting.pdf
14. Bodnar, P. and Grbusic, T. (2020) "Your Climate Disaster Tax Bill Is Growing". In: *The New York Times*, June 23, 2020. New York: The New York Times Company. Available from: https://www.nytimes.com/2020/06/23/opinion/climate-change-financial-disaster.html
15. Froment, R. and Below, R. (2020) "Disaster Year in Review 2019". In: *CRED CRUNCH Issue 58*. Louvain: Centre for Research on the Epidemiology of Disasters (CRED). Available from: https://www.preventionweb.net/publications/view/71642
16. Roach, J. (2020) "Australia wildfire damages and losses to exceed $100 billion, AccuWeather estimates" in *AccuWeather*, 7 January, 2020. Available from: https://www.accuweather.com/en/business/australia-wildfire-economic-damages-and-losses-to-reach-110-billion/657235
17. Network for Greening the Financial System (NGFS) (2019a) "Macroeconomic and Financial Stability Implications of Climate Change". *Technical Supplement to the First Comprehensive Report*. Paris, FR: NGFS Secretariat/Banque de France.
18. Wei, D., Cameron, E., Harris, S., Prattico, E., Scheerder, G., and Zhou, J. (2016) *The Paris Agreement: What It Means for Business*. New York: We Mean Business. Available from: https://www.bsr.org/reports/BSR_WeMeanBusiness_Business_Climate_Paris_Agreement_Implications.pdf
19. For more information see the recommendations of the Biden-Sanders Unity Task Force. Available from: https://joebiden.com/wp-content/uploads/2020/08/UNITY-TASK-FORCE-RECOMMENDATIONS.pdf
20. The Center for Climate and Energy Solutions maps US state-level climate action. Available from: https://www.c2es.org/document/greenhouse-gas-emissions-targets/
21. Myers, S.L. (2020) "China's Pledge to Be Carbon Neutral by 2060: What It Means". In: *The New York Times*, 23 September 2020. New York: The New York Times Company. Available from: https://www.nytimes.com/2020/09/23/world/asia/china-climate-change.html
22. Gerretsen, I. (2020) "UK announces stronger 2030 emissions target, setting the bar for ambition summit" in *Climate Home News*, 3 December 2020. Available from: https://www.climatechangenews.com/2020/12/03/uk-announces-stronger-2030-emissions-target-setting-bar-ambition-summit/
23. Sengupta, S. and Popovich, N. (2019) "More Than 60 Countries Say They'll Zero Out Carbon Emissions. The Catch? They're Not the Big Emitters". In: *The New York Times*, 25 September 2019. New York: The New York Times Company. Available from http://www.nytimes.com/interactive/2019/09/25/climate/un-net-zero-emissions.html

24. A sample of commitments taken by 1600 companies can be accessed through "We Mean Business". Available from: https://www.wemeanbusinesscoalition.org
25. More information available from UNEP FI. https://www.unepfi.org/net-zero-alliance/
26. Systemiq (2020) *The Paris Effect: How the Climate Agreement Is Reshaping the Global Economy*. London: Systemiq. Available from: https://www.systemiq.earth/wp-content/uploads/2020/12/The-Paris-Effect_Full-Report-1.pdf
27. Larson, E., Greig, C., Jenkins, J., Mayfield, E., Pascale, A., Zhang, C., Drossman, J., Williams, R., Pacala S., Socolow, R., Baik, E.J., Birdsey, R., Duke, R., Jones, R., Haley, B., Leslie, E., Paustian, K., and Swan, A. (2020) *Net-Zero America: Potential Pathways, Infrastructure, and Impacts, Interim Report*. Princeton, NJ: Princeton University. Available from: https://environmenthalfcentury.princeton.edu/sites/g/files/toruqf331/files/2020-12/Princeton_NZA_Interim_Report_15_Dec_2020_FINAL.pdf
28. Desjardins, J. (2016) "The Decline of Coal in Three Charts". *Visual Capitalist*, 1 July 2016. Available from: https://www.visualcapitalist.com/decline-of-coal-three-charts/
29. Kuykendall, T. (2020) US coal stocks continue sharp decline into 2020. S&P Global Market Intelligence. Available from: https://www.spglobal.com/marketintelligence/en/news-insights/latest-news-headlines/us-coal-stocks-continue-sharp-decline-into-2020-56940158
30. Shearer, C. (2020) Guest post: How plans for new coal are changing around the world. CarbonBrief. Available from: https://www.carbonbrief.org/guest-post-how-plans-for-new-coal-are-changing-around-the-world
31. Gabbatiss, J. (2019) UK primary energy use in 2018 was the lowest in half a century. Available from: https://www.carbonbrief.org/analysis-uk-primary-energy-use-in-2018-was-the-lowest-in-half-a-century
32. Mendelevitch, R., Hauenstein, C. and Holz, F. (2019) "The Death Spiral of Coal in the US: Will Changes in US Policy Turn the Tide?". *Climate Policy* 19(10): 1310–1324. doi: 10.1080/14693062.2019.1641462. Available from: https://www.tandfonline.com/doi/full/10.1080/14693062.2019.1641462
33. Peña, D. (2020) "In America's Oil Country, Men Losing Their Jobs Are Suffering in Silence". In: *The Guardian*, 20 October 2020. Manchester: The Guardian. Available from: https://www.theguardian.com/us-news/2020/oct/20/unemployment-men-lose-job-self-esteem
34. Levin, K. and Davis, C. (2019) *What Does "Net-Zero Emissions" Mean? 8 Common Questions, Answered*. Washington, DC: World Resources Institute (WRI). Available from: https://www.wri.org/blog/2019/09/what-does-net-zero-emissions-mean-6-common-questions-answered?sfns=mo
35. Raskin, S. (2020) "Why Is the Fed Spending So Much Money on a Dying Industry?". In: *The New York Times*. 28 May 2020. New York: The New York Times Company. Available from: https://www.nytimes.com/2020/05/28/opinion/fed-fossil-fuels.html?referringSource=articleShare
36. Rainforest Action Network (RAN) (2020) *Banking on Climate Change: Fossil Fuel Finance Report 2020*. Available from: http://priceofoil.org/content/uploads/2020/03/Banking_on_Climate_Change_2020.pdf
37. Mercure, J.F., Pollitt, H., Viñuales, J.E., Edwards, N.R., Holden, P.B., Chewpreecha, U., and Knobloch, F. (2018). "Macroeconomic Impact of Stranded Fossil Fuel Assets". *Nature Climate Change* 8(7): 588–593. doi:10.1038/s41558-018-0182-1. Available from: https://www.nature.com/articles/s41558-018-0182-1?WT.ec_id=NCLIMATE-201807&spMailingID=56915405&spUserID=ODE0MzAwNjg5MAS2&spJobID=1440158046&spReportId=MTQ0MDE1ODA0NgS2

38. International Energy Agency (IEA) (2020). *The Oil and Gas Industry in Energy Transitions: Insights from IEA Analysis*. Paris, FR: International Energy Agency. Available from: https://www.iea.org/reports/the-oil-and-gas-industry-in-energy-transitions
39. World Economic Forum (WEF) (2020) *The Global Risks Report 2020*. Geneva: WEF. Available from: http://www3.weforum.org/docs/WEF_Global_Risk_Report_2020.pdf
40. Christopher, M.L. (1992) *Logistics and Supply Chain Management*. London: Pitman Publishing.
41. Hendricks, K.B. and Singhal, V.R. 2003. "The Effect of Supply Chain Glitches on Shareholder Wealth". *Journal of Operations Management 21*: 501–522. http://citeseerx.ist.psu.edu/viewdoc/download?doi=10.1.1.463.4120&rep=rep1&type=pdf[
42. Sheffi, Y. (2007) *The Resilient Enterprise: Overcoming Vulnerability for Competitive Advantage*. Cambridge, MA: The MIT Press.
43. Sheffi, Y. (2017) *The Power of Resilience: How the Best Companies Manage the Unexpected*. Cambridge, MA: The MIT Press.
44. Standard Chartered (2015) *Global Supply Chains: New Directions. Global Research Special Report*. Singapore: Standard Chartered Research. https://www.sc.com/en/media/press-release/global-supply-chains-new-directions
45. Sabin Center for Climate Change Law, Columbia Law School (Sabin Center) (2020) *Climate Litigation Database*. New York, NY: Columbia University. Retrieved from www.climatecasechart.com
46. The Economist (2017) "Climate-change lawsuits". *The Economist,* 2 November, 2017. https://www.economist.com/international/2017/11/02/climate-change-lawsuits
47. Vizcarra, H.V. (2020) "The Reasonable Investor and Climate-Related Information: Changing Expectations for Financial Disclosures". *The Environmental Law Reporter 50*(2): 10106. Available from: http://eelp.law.harvard.edu/wp-content/uploads/50.10106.pdf
48. Gary, S.N. (2019) "Best Interests in the Long Term: Fiduciary Duties and ESG Integration". *University of Colorado Law Review 90*: 731. Available from: https://papers.ssrn.com/sol3/papers.cfm?abstract_id=3149856
49. Keenan, J.M. (2018) *Climate Adaptation Finance and Investment in California*. London, UK: Routledge.
50. Medland, D. (2015) "A 2C World Might Be Insurable, a 4C World Certainly Would Not Be". *Forbes*, published on 26 May 2015. https://www.forbes.com/sites/dinamedland/2015/05/26/a-2c-world-might-be-insurable-a-4c-world-certainly-would-not-be/?sh=58ebca622de0
51. Farmer, L. (2016) "Pension Fund Takes Unprecedented Climate Change Action". *IEN. 17 March 2016*. Andover: The Crane Institute of Sustainability. http://www.governing.com/topics/finance/gov-climate-change-pension-calpers.html
52. Kerber, R. (2017) "Exclusive: BlackRock Vows New Pressure on Climate, Board Diversity. *Reuters Business News, 13 March 2017*. http://www.reuters.com/article/us-blackrock-climate-exclusive-idUSKBN16K0CR
53. Cameron, E., Lemos, M., and Winterberg, S. (2018) *Climate and Inclusive Economy: The Business Case for Action*. San Francisco: BSR. https://www.bsr.org/en/our-insights/report-view/climate-inclusive-economy-the-business-case-for-action
54. Stanley, M. (2017) *Sustainable Signals: New Data from the Individual Investor. Institute for Sustainable Investing*. New York: Morgan Stanley & Co. LLC and Morgan Stanley Smith Barney LLC. https://www.morganstanley.com/pub/content/dam/msdotcom/ideas/sustainable-signals/pdf/Sustainable_Signals_Whitepaper.pdf

55. Avory, B., Cameron, E., Erickson, C., and Fresia, P. (2015) *Climate Resilience and the Role of the Private Sector in Thailand: Case Studies on Building Resilience and Adaptive Capacity.* Hong Kong: BSR. http://www.bsr.org/reports/BSR_Climate_Resilience_Role_Private_Sector_Thailand_2015.pdf
56. Elkington, J. (2020) *Green Swans: The Coming Boom in Regenerative Capitalism.* New York: Fast Company Press.
57. Cameron, E., Erickson, C., and Schuchard, R. (2014) *Business in a Climate-Constrained World: Catalyzing a Climate-Resilient Future through the Power of the Private Sector.* New York: BSR. https://www.bsr.org/reports/BSR_Business_in_a_Climate_Constrained_World_Report.pdf
58. Wong, J.C. (2018) "Hospitals face critical shortage of IV bags due to Puerto Rico hurricane". In: *The Guardian*, 10 January, 2018. Manchester: The Guardian. Available from: https://www.theguardian.com/us-news/2018/jan/10/hurricane-maria-puerto-rico-iv-bag-shortage-hospitals
59. MacWilliams, J.J., La Monaca, S., and Kobus, J. (2019, August) "PG&E: Market and Policy Perspectives on the First Climate Change Bankruptcy". *Center on Global Energy Policy.* New York, NY: Columbia University, School of International and Public Affairs. Available from: https://www.energypolicy.columbia.edu/research/report/pge-market-and-policy-perspectives-first-climate-change-bankruptcy
60. Cameron, E. et al. (2019) *Companies and Climate Resilience: Mobilizing the power of the Private Sector to Address Climate Risks.* Den Haag, Netherlands: Red Cross Red Crescent Climate Centre. Available from: https://static1.squarespace.com/static/59486a61e58c62bbb5b3b085/t/5d41ec40262f7500014c8ed3/1564601409741/RCCC+Private+sector+working+paper.pdf
61. Mukherji, B. (2016) "Coal's Collapse Scorches Miners' Profit Margins". In: *The Wall Street Journal*, 23 February 2016. New York: The Wall Street Journal. Available from: https://www.wsj.com/articles/coals-collapse-scorches-miners-profit-margins-1456228748
62. Greil, A. (2011) "Swiss Re sees Thai Floods Costs at $600 Million". *The Wall Street Journal.* Published on 6 December 2011. https://www.wsj.com/articles/SB10001424052970204770404577081703295323634

2
OVERCOMING SYSTEMIC BARRIERS
Shifting to a new climate economy

As the previous chapters demonstrated, there is both unprecedented risk and opportunity related to the impacts of climate change. The response from the private sector shows that climate action is aligned with business objectives too. At the same time, we see signals of change outside the realm of business action.

- Youth activism, represented by what has become the iconic Greta Thunberg, has mobilized youth worldwide, and it has raised awareness in the mainstream too. What we are witnessing is not the beginning of youth activism – teenagers and young adults in Africa got involved as early as 2006 at COP12 when they launched the "African Youth Initiative on Climate Change." But the movement has grown, with an estimated 1.6 million young people from 125 countries protesting in March 2019. Thunberg's – *Time Magazine*'s person of the year 2019 – collected speeches became a number one NY Times Bestseller, and she led a worldwide mobilization of climate marches in 2019 that led approximately seven million people to the world's largest protest in history. In addition, other forms of activism are growing, exemplified by the organization Extinction Rebellion, also known as "XR," a citizen movement that resorts to non-violent civil disobedience amongst other tactics to put pressure on decision-makers.
- Climate litigation is increasing and is used as a powerful tool to influence policy and raise the ambition of governments as well as to bring about climate justice through compensation from major emitters. Investors, activist shareholders, cities, states, youth are pursuing claims in 28 countries, and the focus is shifting from governments to high-emitting companies. While more than three-quarters of cases are recorded in the United States, litigation in low- and middle-income countries is also increasing in number and in significance. Litigants bring to light the inadequacy or injustice of

climate action and inaction with a view to protecting individual and collective human rights and to raise ambition on legal frameworks (Burger and Gundlach, 2017, p. 40; Setzer and Byrnes, 2019, p. 1).
- Curricula at all levels of education are starting to include courses about climate change, albeit timidly. Environmental education is not new and has been incorporated at country levels since the 1970s in countries such as the United Kingdom, Denmark, or China. What is new is a focus on climate change and sustainability – with Italy, for instance, becoming the first country to making climate change education mandatory in schools in 2019 – as well as more common integration into higher education, and in particular business schools. Top-rated business schools, such as MIT's Sloan School of Management, Duke's Fuqua School of Business, and Yale's School of Management, offer dual programs that cover sustainability and climate. The University of Vermont went so far as to replace in traditional MBA program wholesale in 2012 with an MBA in Sustainable Innovation. These trends are encouraging, insofar as they are preparing the next generation to address climate change, but it is important that climate change education remains exceptional at this stage and far from being the norm.
- Divestment campaigns have exerted pressure on universities to divest from financial products tied to fossil fuels, which impacts the sector and stigmatizes it. The campaign picked up momentum in 2011 by pressuring less than ten US institutions and within a year had reached about 50 campuses. The momentum has grown exponentially and as of March 2021, the number of institutions has grown to 1,312 beyond universities worldwide with a total combined asset value of US$14.56 trillion (Fossil Free: Divestment, 2019).
- Momentum is also growing in politics with politicians campaigning on explicitly climate-focused platforms. The Green New Deal in the United States is a congressional resolution that proposes a package of legislation that addresses climate change and economic inequality within a framework of climate justice. Climate change politics are complex in the United States, and the emergence of this proposal in 2019 signals a shift to politicians recognizing that climate change is not an electoral liability, even using a turn of phrase that echoes one of the most popular US policy packages that is arguably at the root of the country's 20th-century prosperity.
- As political platforms are shifting, citizens are also shifting their expectations as employees. Arising within an overall trend of employee concern for environmental and social concern, focus on climate is also growing here, with, for instance, 76% of millennials willing to prioritize those in favor of pay (Cone Communication, 2016, p. 1). Studies have shown that climate change and environmental issues are the top concern of employees in the United Kingdom in 2019 and that in the same year, these issues jumped in importance by 52% globally amongst that group (Peakon, 2020, p. 6). Younger employees from the so-called "generation Z" led this shift, with a 128% surge. In addition, an increase in employee activism has demonstrated

that employees are going beyond concern to action: in April 2019, 4,200 Amazon employees pressured leadership – specifically CEO Jeff Bezos – to make genuine and impactful commitments on climate change across all operations rather than adopting piecemeal measures with reduced impact. Results were encouraging, with the company announcing the "Shipment Zero" measure shortly thereafter (Clark, 2019). They were less encouraging when Amazon allegedly threatened to fire two activist employees involved with the movement (Millman, 2020).

- Companies are communicating much more directly about climate action than ever before. Corporate communications about climate have not always accompanied climate action itself for a variety of reasons: some companies were not always confident enough about results to submit them to public scrutiny, others did not want to bring attention to their environmental performance as a whole, others yet believed that it was inconsistent with other aspects of the communications and marketing. One telling example of the latter case is Chanel, the legendary French luxury fashion house, which announced a very ambitious climate plan in March 2020 for the entirety of its activities (Chanel, 2020). The company had already been engaged in climate action but had never communicated about it to the general public. What changed was the realization that climate action and luxury are not a paradoxical union, a view that was widespread in the sector, and that, indeed, there are leadership opportunities in taking a stand on these issues. Kering had already attempted to take on this leadership role for the sector since publishing its Environmental Profit & Loss (EP&L) methodology (Kering, 2019), paving the way for the notoriously silent Chanel to make this big splash. It is unclear whether companies such as Chanel are being reactive to new consumer trends or whether they are shaping these trends, but what is clear is that their approach to corporate communications has shifted drastically.

Given what looks like a massive change in the public's approach to climate change, we could expect this to translate into equally impressive changes in our global climate performance. And yet the UNEP Emissions Gap report for 2019 showed once again the gap between actions parties have committed to under their Nationally Determined Contributions (NDCs) and what the best available science tells us is necessary to keep the world below 2 degrees Celsius (°C), let alone the more ambitious 1.5°C target (UNEP, 2019, p. XIII). In 2018, the rate of increase of energy demand increased by more than twice since 2010 (IEA, 2019); at the time of writing, the concentration of CO_2 in the atmosphere is the highest it has been in 63 years at 416.46 ppm (CO2.Earth, 2021); in 2018, global greenhouse gas (GHG) emissions increased compared to 2017 (Crippa et al., 2019, p. 5; IEA, 2020); ocean heat reached record highs that year too (WMO, 2019, p. 5).

In Q1 2020, CO_2 emissions declined the most in the regions that suffered the earliest and largest impacts of COVID-19: China (–8%), the European Union

(−8%), and the United States (−9%). At the time of writing, it is projected that global CO_2 emissions will decline even more rapidly across the remaining nine months of the year, to reach 30.6 Gt for 2020, almost 8% lower than in 2019. This would be the lowest level since 2010. Such a reduction would be the largest ever, "six times larger than the previous record reduction of 0.4 Gt in 2009 due to the financial crisis and twice as large as the combined total of all previous reductions since the end of World War II." This, of course, is due to a slowdown in energy demand, in manufacturing and consumption, and in transport, due to stringent restrictions on movement, in most major economies. It is expected that these records will not prove to set a trend but to reflect drastic – and temporary – changes to our consumption and transport patterns.

Are we witnessing a genuine change, then, or merely a change in attitudes that is not in step with the requirements of the climate crisis both in terms of urgency and in terms of scale? This chapter will offer an analysis of what dynamics have stood in the way of the kind of climate action we need. Given that we have evidence of political will at the local, national, and international level, that the public is not only ready for climate action but is demanding it fervently, that companies and leaders are becoming more literate in the climate crisis, how come indicators are not signaling progress toward the transformation we need?

In what follows, we will present a brief analysis of prevailing theories about these barriers. The point will not be to offer in-depth critiques of these views but to chart the landscape for the reader, offering suggestions for further reading and synthesizing the main views available to build a foundation for moving beyond them.

(What have become) standard theories of the barriers to climate action

Decades of climate action, even if they have followed a path of relative progress, have been stymied by a variety of factors, some of which are structural and proper to our economic systems. Most of the critiques leveled by prominent voices in this debate rely on the observation that features capitalism, specifically, is inconsistent with the systems that would allow for a decarbonized economy. They are all somehow related, in the sense that they all come from a similar conceptual background, which briefly stated holds that our current system does not account for human well-being adequately. In this section, we will provide an overview of those critiques and a short elaboration of this background, with the aim of developing a series of actionable principles for companies throughout the book.

Extractivism

Extractivism describes the processes of extracting resources to sell on the global market through exportation as well as the economic system that sustains these exchanges. The origins of extractivism date back to the Triangular Slave Trade in the late 16th century, where timber, precious metals and stones, raw materials from agriculture, and labor were extracted and then traded through export. The historical path from the Slave Trade to colonization is well known, characterized as it was by production taking place in the colonies to be imported by colonizers through trade. The extraction of raw materials and of labor by colonizers was the basis for much economic growth in the colonizing countries and indeed developed beyond formal decolonization into free-market ideologies that emphasize the export of raw materials from resource-rich countries in the south by state-owned enterprises or by multinational corporations as the pathway to growth-oriented development. Furthermore, as technology has advanced, fossil fuels such as oil, natural gas, and coal have become significant parts of the extraction trade (Acosta, 2013, p. 62; Bebbington and Bebbington, 2012, pp. 17–37; Bunker, 1984, pp. 1017–1064; Harvey, 2007, pp. 22–44; Malm, 2018; Mitchell, 2011).

The environmental consequences of extractivism, which depletes or damages the sources of the commodities it extracts, are clear: disturbance or outright destruction of ecosystems such as soil depletion or ocean heating and acidification, loss of biodiversity, significant GHG emissions (Böhm and Misoczky, 2015, p. 3; Raftopoulos, 2017, p. 394). Extractivism also has social and political impacts by affecting local and indigenous populations, cementing power imbalances between countries or regions, and often leading to corruption, lack of transparency, and poor governance (Acosta, 2013, p. 64). Even without going into the historical scars of the Slave Trade and of colonialism, the presently observable impacts of extractive industries and the global market they thrive on suggests that as long as companies maintain the view of natural resources, especially fossil fuels, as infinitely extractable – which they are not – and hence as growth as infinitely possible, they will not consider alternative business or economic models.

The most comprehensive measurement of the impacts of extractive industries to date, the Global Resources Outlook published by the UN Environment Program and the International Resource Panel in 2019 (Oberle, 2019, p. 60), calculates that 53% of the world's carbon emissions and more than 80% of the global biodiversity loss can be attributed to extractive industries in mining and farming. The same report points out that the rate of extraction has tripled since 1970, while the population has increased at a rate of two only. On the one hand, this has contributed to the rise in certain financial indicators such as GDP; on the other, it highlights how deeply dependent our global economy and our global survival, understood as our needs of food, shelter, clothing that we produce thanks to extracted raw materials, are on extraction and extractivism. The environmental cost and sustainability into the future of these patterns are clear.

Driven by the underlying view that companies' goal is to "grow" (revenue, profit, or dividends, and so on), corporate executives will tend to prioritize whatever allows for this growth to occur as much as possible, that is, in the longest term or at the highest rate. Through extractivism, including extractivism of labor, the illusion that resources are infinite is maintained. Indeed, resources have been extracted more or less continuously for about 500 years, although they are depleting. This has led to the destruction of the very ecological conditions of the possibility of capitalism, and this is on top of the legacy of injustice caused by the acquisition of these resources via expropriation, violent plunder, or colonial dispossession. With ecological, social, and political breakdowns well documented and rife across the planet, this illusion becomes less and less tenable. The critique of extractivism, then, consists in pointing out, on the one hand, that extractivism has devastating consequences for climate by its methods, and because of the glut of fossil fuels it unleashes onto the market, but also because it is tied to a vision of unconstrained growth that inherently disregards these impacts.

The myth of the "private sector"

This section may come as a surprise in a book about corporate leadership. Indeed, it would be reasonable to assume that its arguments and frameworks rest on the view that the "private sector" is not a myth and that it has the power to achieve some of humanity's most important goals, such as mitigating the impacts of climate change and curbing its rise. So when we refer to the "myth of the private sector," here we do not mean that the private sector or its potential to bring about change is not real and needs to be debunked wholesale. Rather, we are bringing attention here to some interpretations of the role of the private sector in society and in solving societal problems.

One of the most prominent exponents of these myths, Mariana Mazzucato, argues that one of them, and possibly the most damaging, is that private companies draw their power to contribute to society solely from "individual genius," "entrepreneurship," or being unfettered by regulation, taxes, or labor law. According to Mazzucato, not only is this factually inaccurate, but it also stands in the way of the most promising solutions to pressing issues of collective survival or thriving. It is inaccurate because, as she demonstrates, what makes private-sector success possible is public investment in education, research, and innovation by funding public institutions but also via strategic procurement and by grant-making to private organizations. Google's algorithmic search engine was developed, for instance, thanks to a grant from the US National Science Foundation; Tesla was struggling to secure investment when it received a $465 million loan from the US Department of Energy. Indeed, three of Elon Musk's companies – Tesla, SolarCity, and SpaceX – which are hailed as tech companies with the potential to "save" us and as the fruit of Musk's own entrepreneurial and engineering genius, received a total of nearly $5 billion in public support. She further shows that every piece of technology needed to make the iPhone, from the internet to touch screens, was the result of publicly funded or publicly developed research and innovation programs. Mazzucato goes on to show that without early risk-taking and investment into technologies by the public sector, the development of aviation, nuclear energy, computers, nanotechnology, biotechnology, and the internet would not have occurred as they have. Companies in turn would not have benefited from those technologies to develop their products and generate revenue. It is not that companies do not rely on their own skills and competencies; it is that these do not occur in a vacuum. Instead, they occur in an environment of public–private collaboration where, more specifically, the public sector enables and empowers the private sector to make new goods (Mazzucato, 2011; Mazzucato, 2018b).

The myth she exposes in this way is also shown to be a barrier to progress for several reasons. One of the main issues is that companies that benefit from public investment for the development of their own technologies and hence for their own revenue do not reinvest it back into further research and development or training. Indeed, Mazzucato shows that between 2003 and 2013, S&P 500 companies used more than half of their earnings to buy back shares to boost stock prices. Pfizer spent $139 billion on buying back shares, and Apple, between 2012 and 2018, that is, after Job's tenure, spent almost one trillion dollars on share buybacks. Buying back shares means higher dividends for shareholders, who of course are then free to spend that money as they please – on cars, real estate, yachts, college admissions, or more unlikely, research and training for workers (Lazonick and Mazzucato, 2013, pp. 1093–1128; Lazonick, Mazzucato and Tulum, 2013, pp. 249–267; Mazzucato, 2018b).

It is only once we realize this and design policies accordingly, says Mazzucato, that we are going to find the next wave of innovation that we need to tackle urgent challenges such as climate change. What is more, by making them accessible and widespread, we will be able to reach the scale, the places, and the people we need to reach most. She models her proposed solution on the space program that leads to the Moon landing in 1969. The US government invested $26 billion between 1960 and 1972 in the Apollo Program that achieved this. Notably, what made this possible was not only innovation in aeronautics but also in textiles, nutrition, electronics, medicine, leading to hundreds of products that were developed and commercialized as well as other space projects. The solution to the climate crisis is likely to follow a similar model of public investment into R&D with the outflow of innovation into many sectors and with many results that may or may not be related to climate. In short, and to use Mazzucato's phrase, we need mission-oriented policies that will harness the power of companies and of the market since without framing policies, they alone will not yield the solutions we seek (Mazzucato, 2018a, pp. 803–815; Mazzucato, 2018b).

GDP fetishism

Uniting these critiques is their aim at a widespread and deeply held view about what the economy should be and hence what we should do to make it so. Specifically, the view, common since at least the middle of the last century, that the Gross Domestic Product (GDP) is the best and only indicator of wealth and that growth is what makes it healthy, is held to be a barrier to the kinds of systemic changes that would make necessary climate action possible.

GDP is a number that measures the sum of the market values, that is, of prices, of all final goods and services produced in a country during a period of time. Clearly, there are many other notions that deserve to be unpacked in this definition, such as value and price, but they go beyond the scope of our purposes (Callen, 2020). What matters for us here is that GDP is taken to be a global measure of economic growth – and, over time, of the increase in the amount of goods and services produced over a period of time – and hence of wealth and prosperity. If more goods and services are produced, the growth rate is high, increasing the GDP, which means a country becomes more prosperous, according to the view. A simpler way of putting it is that the more we produce and hence consume, the better off we are as an economy and a country, where "better off" here includes all senses of "well off." Consequently, the economy as a whole should be geared toward increasing GDP, in other words, toward producing and hence consuming more goods and services.

It is this view, the critics hold, that prevents us from addressing climate change at the scale and pace that we need to – notwithstanding significant efforts and opportunities such as the ones we outlined in Chapter 2. There are many versions of this line of objection, and they are not only related to climate and environmental concerns. To understand them, it is necessary to understand that GDP does not discriminate between the types of goods and services, the value of which is counted. So, for instance, it measures equally the production of one US$5,000 pair of shoes and one hundred $50 pairs. Merely intuitively, we can tell that both scenarios paint pictures of very different economies and countries. Moreover, even with growth as the goal, GDP does not tell us whether we should produce more of the expensive pairs or more of the cheaper pairs – all we can deduce is that we should produce, and hence consume, more *simpliciter* (Desai, 2018; Kapoor and Debroy, 2019; Talberth, 2010; The Shift Project, 2012).

Critics argue that this orthodoxy is the reason we find ourselves in the climate predicament we currently face: production at all costs, in total disregard of environmental and social impacts, has led to a blind increase in resource use and energy consumption, and destructive production and distribution methods (Smil, 2019; Jackson, 2017). In addition – and crucially for the objectives of this book – it is also the reason we find ourselves falling short of the impact we are seeking with climate action. Climate action requires that we produce certain goods and services and not others and that we produce them in particular ways. Usually, seeking growth in itself gives us no indication as to what these should be, and when it does, it is more likely that that search will best be served by continuing to produce high-emitting goods and services in high-emitting ways.

Even within mainstream economics, the hegemony of growth as it is captured by GDP is challenged. Nobel Prize winner Joseph Stiglitz argues that the GDP is outmoded since it fails to take into account inequality and environmental degradation, and is a vocal advocate of the "Beyond GDP measurement" agenda that includes other indicators that reflect human well-being more accurately (Stiglitz, 2018). Harvard economics and philosophy professor Amartya Sen has also offered an alternative vision to GDP measurement that would include measures of health, freedom, and education for a better record of what counts toward human well-being. Together, Stiglitz and Sen published a foundational critique of GDP and spurred this important line of thought amongst economists and beyond (Stiglitz, Sen and Fitoussi, 2010). The two winners of the 2019 Nobel Prize winners, Abhijit Banerjee and Esther Duflo, have also pointed out that larger GDP does not simply translate into higher human well-being. They also point out that in advanced countries, the pursuit of growth has been counter-productive, contributing to a rise in inequality, mortality rates, and political polarization. Growth has benefits, but it tends to be captured by those that already benefit from the existing distribution of wealth.

The recent laureates do note that economic growth can have benefits, especially in countries where poverty is rife. Indeed since 1990, the number of people living above the World Bank's threshold of extreme poverty of $1.90 a day has fallen from almost two billion to approximately seven hundred million, and "in addition to increasing people's income, steadily expanding GDPs have allowed governments (and others) to spend more on schools, hospitals, medicines, and income transfers to the poor" (Banerjee and Duflo, 2020). A simple rejection of all growth indicators, then, would mean foregoing measuring aspects of development and that by consequence, what is perhaps a better way forward is a hybrid approach.

What is important is not the number provided by GDP measurements, this suggests, but a broader swathe of indicators that also account for wealth distribution on the one hand and on qualitative aspects of what is produced and how in a country over a period of time (OECD, 2015).

The upshot of accuracy would of course be a benefit of this approach, but in the context of climate action, jettisoning the blind pursuit of ever-increasing production at all costs would more importantly liberate corporate decision-making makers from revenue, sales, and profits as the north stars of business performance. It could allow, depending on the alternative, for the integration of the value of environmental factors and of minimizing impacts and could steer the economy wholesale toward low-emitting modes of production.

"UNCOUPLING": GDP, ENERGY DEMAND, AND EMISSIONS

One corollary of setting GDP as the standard of prosperity is that economic growth is taken to imply increases in energy demand and that conversely reducing energy demand will lead to slowing down growth (Sharma, Smeets and Tryggestad, 2019).

The IEA has published data on energy-related CO_2 emissions for almost 50 years and emissions have increased all but four times in that period. Three of those coincided with crises in the global economy, occurring in the 1980s, 1992, and 2009. It is further projecting, at the time of writing, that 2020 will follow this pattern: one of the impacts of the COVID 19-pandemic will be, in addition to historical rates of GDP contraction, that energy demand will decrease by 5–10%, with higher rates to be expected the longer lockdowns will last (IEA, 2020).

And yet evidence indicates that this relation between growth and energy demand is changing and we are observing what is called an "uncoupling" of growth and emissions as a result.

- The GDP of the United States has risen by 25% since 2005 (FRED, 2020), while energy-related emissions have fallen by between 10% and 14% in that same period (EPA, 2021; Hausfather, 2017).
- In the UK, GDP per capita grew by 70.7% between 1985 and 2016, while CO_2 emissions fell by 34.2% (Agbugba et al., 2019, p. 8).
- Based on early projections, EU GHG emissions in 2018 amount to a 23% reduction from 1990 levels (EEA, 2020). At the same time, the EU economy has continued to expand in 2019 for the seventh year in a row with GDP increasing in every member state (European Commission, 2019, p. 2).
- Some estimates show that China's GHG emissions dropped for the first time in 2018, between 0.5% and 1.5%, while growth increased at a rate of 6.9% (Jiang, Zhou and Li, 2018, pp. 1–12; Wu et al., 2019, pp. 576–588).
- In 2018, emissions in Australia were 14% below the 2007 peak and 11.7% percent below 2005 emission levels (Australian Government, 2019, p. 3). Meanwhile, Australia's GDP has grown consistently since 1990 (Country Economy, 2020).

The longer the time horizon over which these trends are observed, the more conclusive results will be and the stronger the hypothesis of uncoupling. These parallel trends, however, together with explanatory factors, very strongly suggest that this is not accidental or cyclical but rather indicative of a shift in the economy.

It is not that the world is less hungry for energy or that fewer people have access to energy. Rather, these four factors support the thesis of uncoupling:

1. As the global economy shifts from industrial to service economies, especially in fast-growing regions like China and India, the energy intensity of GDP falls.
2. Energy efficiency is increasing as a result of technological advances and behavioral changes.
3. Electrification of transport in particular represents a large reduction of emissions and opens up ways to meet energy needs in other areas too through advances in battery technology.
4. The increase of renewables in the share of electricity usage contributes to reducing GHG emissions.

> The energy sector is poised to be transformed between today and 2050 and this will have deep consequences for businesses, who will see their business models challenged and who will also see opportunities for cost savings and efficiency gains. Most of all, these factors explain that climate action around the mitigation of emissions is not opposed to economic growth, on the one hand. On the other hand, the yardstick of economic growth and GDP will become obsolete as a measure of prosperity and progress, revealing that new measures and hence new frameworks will surpass it.

These analyses of the barriers to climate action at scale and pace are all important and help us understand the disconnect between the genuine momentum and opportunity we have highlighted and the "emissions gap" that is also well documented. The aim of the section was to present these and to give the reader a sense of the landscape of these objections and to provide the background for what follows. Next, we will review how these critiques have spurred creative models for transforming the economy.

Some proposed solutions and their limits

Proposals for the kind of transformation necessary to tackle the climate crisis are relatively rife in the literature and tend to find part of their theoretical foundation in the types of critiques we have just reviewed. The main lines of the argument tend to broaden their focus on the economy into broader conceptions of society or nature, casting economic actors with the power to transform both of them. In this section, we will review four frameworks that explain corporate action in the realm of climate and the environment and will assess whether they are able to address the systemic barriers we have identified so far.

Corporate Social Responsibility

Corporate Social Responsibility (CSR) is one of the oldest modern formalizations of extra-financial action by companies, with conceptual frameworks having been developed for well over 50 years (Carroll, 2008). For the most part, and historically, CSR can be understood as the enlightened self-interest of companies: they will take on social and environmental responsibilities insofar as doing so favors the pursuit of their self-interest, usually understood in financial terms.

The most common model of CSR relies on the framework of the triple bottom line, popularized by John Elkington (2004), which shows the relationships between the financial, environmental, and social returns of a company's operations. The phrase "people, planet, and profit" originates from this framework and is meant to capture that a company's goals should be equally prioritizing all three. Expanding traditional accountability concepts, this framework includes areas of action that had prior been considered ancillary and instrumental to business performance rather than indicators of its success.

To justify engaging in CSR, companies advance arguments about (a) generating revenues over the long-term, (b) cost savings, (c) risk management, (d), securing a social license to operate. These all amount to reasons that advance the business objectives of companies in ways that do not fit in other types of decision-making frameworks regulating companies. The profitability of CSR measures has been claimed but not proved irrefutably, though some positive correlations have been documented. Consequently, companies have often used other types of an argument than strictly financial ones to justify their investment in sustainability measures (Gond and Crane, 2010, pp. 677–703; Margolis and Walsh, 2003, pp. 268–305; Orlitzky, Schmidt and Rynes, 2003, pp. 403–441).

This framework guides the vast majority of companies' concrete action on climate and indeed "CSR" is, in most companies, a division unto itself with entire teams deployed to the "people" and "planet" impacts of the company, with others measuring those impacts and translating them into accounting terms or other quantitative and qualitative indicators. Over 90 percent of the world's largest companies by revenue publish detailed annual reports of their CSR activities as of 2017, according to a survey by KPMG (Blasco and King, 2017, p. 9).

Given that it has become somewhat ubiquitous, it is not surprising that CSR has come under fire. Some have objected to the use of CSR to cover for greenwashing or to detract from other unethical practices and poses a reputational risk as a result (Meier and Cassar, 2018); some have argued that CSR triggers "bad" behavior by "freeloaders" by giving the false impression that ethical standards are upheld by the company without a need for personal engagement (Hedblom, Hickman and List, 2019; List and Momemi, 2017); others would yet argue that while the company's self-interest is still the main objective of any action, it is impossible for companies to truly benefit their stakeholders. John Elkington himself observed recently that most CSR programs amount to "change as usual," or the need for companies to engage in initiatives and to measure and report their results as superseding their actual impacts (Elkington, 2020).

For our purposes, however, we do not need to consider all lines of criticism that have been leveled against the CSR framework, nor do we need to outline them in detail. What matters here is that they all converge to show that CSR is embedded within the very system that causes or exacerbated the environmental or social dysfunctions it seeks to rectify. Tethered as it is to existing accountability methodologies, it provides little promise of overcoming them and instead is more likely to cement the prioritization of growth in business decision-making. What is more, CSR departments are historically separate from core business divisions and are embedded in communications, human resources, or public affairs departments and are rarely directly represented in the C-suite. This distance from the centers of decision-making on operations, supply chain, or finance is to the detriment of CSR actions. Relegated as they are to departments that are conventionally held to be less strategic, it is almost impossible for CSR practitioners to transform a company, never mind an entire sector or system.

Regenerative capitalism

This line of thinking rests on the idea that our current economic models atomize society and nature into independent parts and hence create dysfunctions and imbalances, the climate crisis being but one, of perhaps the most disastrous, consequence.

Most vocally propounded by John Elkington, of "triple bottom line" fame (Elkington, 2004), the idea of regenerative capitalism rests on "universal principles and patterns of systemic health and development [that] guide behavior" (Fullerton, 2015, p. 7) in all systems, including the economy, and the idea that they are governed holistically: "everything in the universe is organized into "systems" whose interlinked parts work together in some larger process or pattern." (Fullerton, 2015, p. 6). Once we realize this, the theory goes, we no longer see the task ahead as one of "solving problems" but as building healthy systems holistically. Elkington summarizes the approach in this way:

> Instead of assuming economic efficiency and undifferentiated GDP growth automatically lead to prosperity, actors in a Regenerative Economy understand that long-term economic vitality depends on creating conditions that will unlock the vast potential for true wealth creation that lies dormant in every individual, community, business network, and bioregion. Consequently, instead of viewing moral issues as irrelevant to "rational" economic decision-making, in Regenerative Capitalism, human and moral concerns become central to decision-making, and policymakers view those concerns as critical to the maintenance of a healthy whole. In this sense, Regenerative Capitalism is also a humanist capitalism. (Fullerton, 2015, p. 10)

Clearly, this solution aims to rethink the different logics underpinning business decision-making by challenging the current incarnation of capitalism. By seeing the interconnections between systems, we can not only make them healthier – in the case of climate, this would mean reducing GHG emissions, amongst other consequences – but also create "true wealth" out of them. What is more, this would be done within the constraints of ethics and morality that would respect all beings in those systems.

Elkington has gone on to explain how this shift toward regenerative capitalism can happen. Given at once the urgency of the issue and its complexity, this transformation will occur through breakthroughs. Specifically, he defines them as "green swans," which

> deliver exponential progress in the form of economic, social, and environmental wealth creation. At worst, it achieves this outcome in two

dimensions while holding the third steady. There may be a period of adjustment where one or more dimensions underperform, but the aim is an integrated break-through in all three dimensions. (Elkington, 2020)

Through innovation, we can seize the opportunity to steer the economy toward regenerative capitalism. These breakthroughs will be "exponential," which means they will have ripple effects such as transforming entire systems, and they will be at the right pace and scale for the problems they solve.

This proposal puts a lot of weight on innovation to transform our economy, and we have already evoked some of the complexities about how innovation comes about. Capitalism is deeply embedded in culture, politics, the day-to-day life of every individual on the planet, and hence it is conceivable that several levers be necessary to radically change it. However, there is value in a model for change that is forward-looking, makes place for solutions we have not yet conceived and is holistic in essence. Indeed, this suggests that the ties that capitalism has severed between human well-being or survival and economic value, a disconnect epitomized by the GDP yardstick, can be restored and hence that we may find the right solutions to address environmental and social issues collectively.

This framework is helpful insofar as it recognizes the limits and irreversible risks of the current system. However, there is little guidance from Elkington as to how we may steer innovation toward the right goals, other than averting those risks. While this is a valuable and indeed extremely difficult goal, Elkington's latest framework does not help us design a view of the future so much as a rejection of the present. He suggests that we have a duty toward the future, specifically future people, but falls short of explaining how this would avert a future we wish to avoid and create the one we want, apart from helping us "create system value," which will by definition, or perhaps by postulation, reconnect us to human well-being. In fact, a clue to the kind of future Elkington wants comes from his view that "only business has the power to address these critical challenges at the necessary pace and scale" (p. 148). He certainly thinks it will be a fully transformed kind of business that will itself bring about the future of "green swans" but it will be without "governments and the public sector." The solution, it seems, consists in the latter making way for the companies to redefine value – and to let Elkington help them redefine it as "systemic value." Consequently, it is unclear that the kinds of changes he advocates for will bring about the systemic change – away from "GDP fetishism" and its variants, perhaps – and not, instead, a business-driven, unaccountable, and non-transparent new world order. This would be a genuine concern if the business could indeed operate without governments and the public sector, specifically for purposes of innovation, so we need not worry too much that Elkington's most current vision will come to be, notwithstanding the helpful elements it provides.

Shared value

Setting business in a network of different actors including government and the public sector, the framework of Shared Value takes a different approach than Elkington's. Shared value was introduced by two of the most influential thinkers in business strategy, Michael Porter and Mark Kramer. Michael Porter will no doubt be familiar to anyone who has taken an interest in business strategy as the architect of "Porter's Five Forces" (Porter, 1979). Since 1979, when this framework was first published, Michael Porter has remained influential including by continuing to publish research and by advising companies, governments, and NGOs. With Mark Kramer, he founded a social impact consultancy, FSG Social Impact Advisors in 2000, and more relevant to our current focus, they introduced the concept of Shared Value in 2011 (Porter and Kramer, 2011).

They define shared value as "policies and operating practices that enhance the competitiveness of a company while simultaneously advancing the economic and social conditions in the communities in which it operates. Shared value creation focuses on identifying and expanding the connections between societal and economic progress" (Porter and Kramer, 2011). This has ramifications for business decisions on social and environmental impacts while maintaining the target of generating profits, sales, and revenue. Indeed, the theory is characterized by recognizing that seeking to maximize "shared value" is not only compatible with but also essential to the long-term sustainability of the business including from a financial point of view.

Companies can use shared value as a way to define and design their strategy, which will help them solve social issues profitably. By leveraging the resources and innovation of the private sector to create new solutions to some of society's most pressing issues, companies create a more prosperous environment in which to operate, making business more sustainable and resilient. There are three pillars to achieving shared value:

1. Reconceiving products and markets in ways that better serve society's needs.
2. Redefining productivity in the value chain with an impact on how resources, energy, suppliers, logistics, and labor are accessed and used.
3. Creating and enabling local development by improving skill development and capacity building in the local operating environment.

Accomplishing these relies on an ecosystem of actors and societal conditions that can either enable or hinder shared value-oriented action, and awareness and transformation of these falls under the remit of a given company. For instance, businesses need to foster and participate in multisector coalitions that can advance shared value efforts. As a result, companies achieve collective impact, embedded as they are in complex systems of collaboration, as opposed to individually measured indicators.

What is innovative and promising about this approach is that it embraces a new understanding of the relationships between stakeholders that make up the ecosystem of a company. On top of identifying the *issues* that are traditionally understood to be beyond the remit of business as essential to the long-term survival and potential thriving of business, Share Value also identifies the *ecosystem of actors* involved in transformation as non-optional to achieve systemic change. Indeed, to change a system, it stands to reason that the entire system should be engaged. In this way, Shared Value places business, in partnership with this broad ecosystem, as a powerful actor toward the systemic transformation of the economy.

This view runs up against its own limits, however, in that it is still geared toward generating growth in some way or other. The goal of creating a more prosperous environment in which to operate is by definition geared toward the continued existence and growth of a company by generating revenues, sales, and profits in existing and even new markets. Without challenging this very premise, the framework of Shared Value runs the risk of perpetuating the systems it seeks to rectify. Indeed, it even exposes itself to the critiques of CSR we outlined above, according to which all actions resulting from the framework are reducible entirely to a growth mindset of financial self-interest.

Policy engagement and progressive lobbying

Companies getting involved in politics usually evokes opaque dealings that are meant to advance the financial interests of companies against the common good. Since the 1950s, tobacco companies have notoriously wielded influence in politics and the public sphere by deploying illicit and covert tactics. Corruption, hiding information, lobbying for laws and tax rules all contributed to creating a favorable environment for them to operate and become extremely profitable, all the while knowingly harming the public through addiction, cancer, and other fatal or near-fatal illnesses. Emblematic as it is, the tobacco industry is not the only sector to have engaged in these sorts of practices, and we know that the fossil fuel industry has been guilty of similar deeds, as has finance, agribusiness, or fast fashion (Mayer, 2016; Oreskes and Conway, 2010).

However, the relationship between business and policy need not be rife with illegal and immoral dealings. Within certain principled constraints of transparency, accountability, and legality, business and policymakers can collaborate to bring about the changes necessary to tackle the climate crisis at a pace and scale. We Mean Business, a coalition of organizations that has led to more than 1,600 of the world's largest businesses making more than 2,000 commitments on climate action, has achieved its success by unifying the voice of climate-progressive companies to policymakers. Supported by a theory of change according to which companies can influence global policy and contribute to creating enabling environments for an inclusive transition to a net-zero economy because of their weight in the real economy, these companies have genuine leverage, individually or as a group, to lend support to public officials to raise the ambition of climate policy. Companies are not acting selflessly – and this is why their voices are taken to be authentic – since they are asking for the kind of policy frameworks they need in order to reach their own climate targets, such as better policy around electricity procurement or financial and risk disclosure. This theory of change was manifested in the lead up to COP21, evidenced by the "Business Brief" that was targeted at policymakers and which expressed what progressive business needed out of the Paris Agreement (We Mean Business, 2015). The eight asks contained in the document are reflected in the Paris Agreement, marking a clear success for the business community that was seeking to raise its own ambition on climate.

There are many tactics that will amount to an influence strategy of business on politics, and they will be defined by the moment, the aims, the messenger, and so forth. It differs, crucially, from a communications campaign in that it is devised to have a genuine influence on policy and is based on the real transformation needs of a company or of a sector. Indeed, in the case of We Mean Business strategy around COP21, the asks were based on analysis-driven needs of companies, compatible with reputational, leadership, or communications opportunities for the companies involved.

How does this involvement in policy influence contribute to the systemic changes we need to close the emissions gap? Indeed, companies have already been engaged in this kind of activity and we find ourselves in a predicament where we have come short. First of all, this solution is not presented here as a silver bullet, and secondly, it is not presented as incapable of being improved upon. Corporate involvement in politics is marred with a questionable past and this is due to the lack of political accountability of businesses. Hence for this path to have its intended effects and to avoid backfiring, it is key for it to be embedded within the most transparent methods of influence.

Purpose

In the effort to redefine the goals of companies in ways richer than a GDP mindset can provide, some companies frame their strategy and decision-making thanks to their purpose. As early as 1957, Harvard economist Carl Kaysen wrote that management should see "itself as responsible to stockholders, employees, customers, the general public, and, perhaps most important, the firm itself as an institution. The view was already prevalent amongst business leaders. Indeed, in 1949 the president of General Foods, Clarence Francis, told Congress that he had a "three-way responsibility to the American consumer, to our associates in this business, and to the 68,000 [stockholders in General Foods]. We … would serve (the company's) interests badly by shifting the fruits of the enterprise too heavily toward any one of those groups" (Francis, 1949, cited in Wells, 2012, p. 330). In 1951, the president of Standard Oil of New Jersey Frank Abrams claimed that managers needed "to conduct the affairs of the enterprise in such a way as to maintain an equitable and working balance among the claims of the various directly interested groups—stockholders, employees, customers, and the public at large" (Adams, 1951, cited in Smith, 2011, p. 1). And then Milton Friedman famously told the readers of the New York Times in 1970, that the "social responsibility of business is to increase its profits," (Friedman, 1970) a dictum that has become a fundamental tenet of any business education and has even become conventional wisdom. But it was not always so and currently the notion of purpose is gaining more traction amongst business leaders.

By the dawn of 2020, a number of pronouncements signaled that business had reached an inflection point. In the United Kingdom, the British Academy published *Principles for Purposeful Business* (The British Academy, 2019); in the United States, the Business Roundtable, an association of CEOs, issued its *Statement on the Purpose of a Corporation* (Business Roundtable, 2020); in Switzerland and on a global platform, the World Economic Forum published *The Davos Manifesto 2020: the Universal Purpose of a Company in the Fourth Industrial Revolution* (Schwab, 2019). Larry Fink, Chairman and CEO of the world's largest asset management firm Blackrock outlined how purpose would be integrated into the company's investment strategy in his annual letter to CEOs in January 2019 (Fink, 2019).

All of these most definitely go beyond the definition of purpose as solely aimed at increasing profits and lay out ways to understand how not only to embed companies within the complex fabric of society but also how to serve its needs.

The purpose is explicitly defined in terms that elude financial indicators. Purpose points to what differentiates an organization and what is authentic to it by articulating the need it fulfills in the world. It brings to light fundamental elements of an organization and why it exists – its unchanging reason for being. It can be understood as a company's "north star" that allows it to direct all of its actions in a way that serves the systems and stakeholders (i.e. the world) in which a company operates. It does not have a "purpose beyond purpose," such as generating more favorable economic conditions, though it does have results that contribute to the current economy. It is not tied to current economic models, however, that require for companies to sustain them at all costs, and it is designed to help companies weather deep disruptions and contribute to social resilience in tackling crises. The COVID-19 pandemic, for instance, has spurred a slew of analyses about the purpose and its positive impact on brands, operations, leadership, organizational structure, human resources (Armano, 2020; EY, 2020; Schaninger et al., 2020), and, more importantly, large-scale action on the part of purpose-driven companies (White and Meese, 2020). When companies have to face the existential questions of what and whom they serve, and when circumstances are such that weaknesses and negative impacts on the world are brought under a stark light, purpose signals an underlying constant that may help them bridge the present to the future.

The contribution of purpose to a company's strategy and direction is something to be worked out and that is not prescribed, so to speak, from above. There is no program that follows from the very concept of purpose and what is required in order to work out the implications of a company's purpose is a detailed mapping of that company's operations, culture, and operating environment. Hence purpose alone does not guarantee that a company will tackle systemic issues in general and the climate crisis in general. Like the other frameworks presented here, it is not a silver bullet and indeed has not shown, at least so far, that it alone has the potential to shift the private sector away from the GDP mindset.

CEO activism and leadership

A new phenomenon has emerged in recent years where CEOs have taken stands on issues that are considered to be political – and sometimes even thorny – and made business decisions to reflect these stands. Corporate activism refers to a form of leadership that speaks out on issues that are not strictly speaking part of the core business of a company. Ben Cohen and Jerry Greenfield, who founded ice cream company Ben & Jerry's in 1978, embedded activism into their company from the outset and even today, consumers can find political messages about refugees, the climate, or conservation on their packaging and in the branding of their various product lines. Unilever CEO Paul Polman – who now owns Ban & Jerry's – has also long encouraged CEOs to push for policies to address climate change – before this was a trend, and with so much success that he is a mainstay in any debate surrounding corporate sustainability. Salesforce CEO Mark Benioff has also become illustrative of CEO dedication to issues broader than business strictly understood by holds dinners for fellow CEOs to explain why he views public activism as part of their job (Bort, 2018). Some form groups to take action – as a group of 30 CEOs with operations in the United States did when Trump announced he would exit the Paris Agreement on climate (The B Team, 2017). CEOs of Goldman Sachs, The Walt Disney Company, The Coca Cola Company, Dow Chemicals, and JP Morgan Chase, among others, urged the then president of the United States to reconsider his position. Others have made more punctual and perhaps more visible moves, such as Merck CEO Ken Frazier stepping down from one of President Trump's business advisory councils to communicate his disapproval of his response to racial violence in Charlottesville, Virginia in 2015 (Gelles, 2018). The list of CEOs of some of the largest companies in the world expressing political views keeps growing, as does the list of issues they take a position on, from immigration, to human rights, LGBTQ issues, climate change, racism, gender equality, and gun control.

There are many drivers behind this new form of leadership: a recognition of the power of companies to have an impact on global issues, which correlates with the converse impact they have on them, the fact that political leadership sometimes makes operating environments more volatile and more difficult to navigate, or the fact that expectations on the part of consumers and employees are shifting (Catalyst, 2021, p. 2; Povaddo, 2017, p. 4). Whatever is driving CEOs to change, the view expressed by Bank of America's CEO Brian Moynihan to the *Wall Street Journal* in 2016 that "our jobs as CEOs now include driving what we think is right, it's not exactly political activism, but it is action on issues beyond business" (Langley, 2016) is gaining traction amongst corporate leaders.

Research on the impact of this type of leadership is still nascent and it is unclear to what extent CEO positions influence public discourse, policy, or government officials (Chatterji and Toffel, 2019). It is further unclear what impact this has on their companies either for consumers of their brands or for employees in their labor force.

While more research is essential in getting a firmer grasp on whether CEO activism has the potential to overcome barriers to systemic change and specifically the barriers that block climate action, we can confidently put forward the hypothesis that the capacity of CEOs of public companies to bring about this kind of change is directly dependent on their duty to shareholders. As long as CEOs are tied to quarterly calendars for financial reporting and as long as they are bound to redistribute dividends to shareholders, their decision-making will be guided in priority by these two imperatives. No doubt if taking a political stand aligns with making good on these, we can expect more CEOs to be vocal.

Moreover, it is unclear whether they represent a genuine challenge to the status quo or whether they are inherently a part of it. Given public expectations on certain issues, the more skeptical amongst us would observe that CEO activism can become an instrument of PR – PR firms have indeed built entire practices around it (Chatterji and Toffel, 2018); others will note that some CEOs have taken stands that go against values of equality and fairness (Vasilogrambros, 2014); other still will note that CEO activism remains a fairly rare phenomenon that only a few CEOs with strong reputations and networks can afford. Finally, most will point out that actions speak louder than words and that a company ought to be evaluated on its performance and real-world impact and not only on the views of an individual, even if that individual sits at its helm. In sum, while the case remains open as to whether this kind of activism will be a necessary condition for tackling the climate crisis, it is clear that it will not be sufficient.

Genuine efforts are underway to tackle societal and systemic issues, including climate change. That these efforts are insufficient is not a blemish to the intentions of those who carry them out or a sign that those efforts are led in error, misguidedness, or incompetence. Rather, that we are collectively falling short is a signal that the solutions offered are not suited to the scale or type of problem we are dealing with. Misdiagnosing the climate predicament as a problem of means-end resolution, or even as a problem of shifting mindsets in companies is to set ourselves up to fall short. The problem has deeper roots that define and constrain the economic system within which corporate actors can tackle climate and seize the opportunity of a new economy in a sustainable way. And that is why, as this chapter will have shown, there is no silver bullet for climate action. The necessary changes are diverse and offer a portfolio of solutions that will be tailored to each unique company. The transformation will not happen overnight and wholesale: it will happen by adopting a series of changes that are all oriented toward shifting to a new climate economy and away from an infinite-growth mindset that is incompatible with it.

References

Acosta, A. (2013) 'Extractivism and Neoextractism: Two Sides of the Same Curse' in Lang, M. and Mokrani, D. (eds.) *Beyond Development Alternative Visions from Latin America* [Online]. Available at: https://www.tni.org/files/download/beyonddevelopment_complete.pdf (Accessed: 20 March 2021)

Agbugba et al. (2019) *The Decoupling of Economic Growth from Carbon Emissions: UK Evidence.* Available at: https://www.ons.gov.uk/economy/nationalaccounts/uksectoraccounts/compendium/economicreview/october2019/thedecouplingofeconomicgrowthfromcarbonemissionsukevidence (Accessed: 20 March 2021)

Armano, D. (2020) 'How the Pandemic is Pressure-Testing a Brand's Purpose,' *Edelman*, 30 March [Online]. Available at: https://www.edelman.com/covid-19/perspectives/testing-brands-purpose (Accessed: 20 March 2021)

Australian Government Department of the Environment and Energy (2019) *Quarterly Update of Australia's National Greenhouse Gas Inventory: March 2019.* Available at: https://www.industry.gov.au/sites/default/files/2020-07/nggi-quarterly-update-mar-2019.pdf (Accessed: 20 March 2021)

Banerjee, A.V. and Duflo, E. (2020) 'How Poverty Ends,' *Foreign Affairs*, January–February [Online]. Available at: https://www.foreignaffairs.com/articles/2019-12-03/how-poverty-ends (Accessed: 20 March 2021)

Bebbington, D.H. and Bebbington, A. (2012) 'Post-What? Extractive Industries, Narratives of Development, and Socio-Environmental Disputes across the (Ostensibly Changing) Andean Region' in Haarstad, H. (ed.) *New Political Spaces in Latin American Natural Resource Governance* [Online]. Available at: https://link.springer.com/chapter/10.1057/9781137073723_2 (Accessed: 20 March 2021)

Blasco, J.S. and King, A. (2017) *The Road Ahead. The KPMG Survey of Corporate Responsibility Reporting 2017.* Available at: https://assets.kpmg/content/dam/kpmg/xx/pdf/2017/10/kpmg-survey-of-corporate-responsibility-reporting-2017.pdf (Accessed: 20 March 2021)

Böhm, S. and Misoczky, M.C. (2015) 'Environment, Extractivism and the Delusions of Nature as Capital' in Mir, R., Willmott, H. and Greenwood, M. (eds.) *The Routledge Companion to Philosophy in Organization Studies.* London: Routledge [online]. Available at: https://www.researchgate.net/publication/301548552_Environment_extractivism_and_the_delusions_of_nature_as_capital (Accessed: 10 September 2016)

Bort, J. (2018) 'How Salesforce's Marc Benioff Has Been Pushing Other CEOs to Do the Right Thing, Even When the Government Won't,' *Business Insider*, 7 April [Online]. Available at: https://www.businessinsider.com/how-salesforce-ceo-pushing-other-ceos-to-do-the-right-thing-2018-4?IR=T (Accessed: 20 March 2021)

Bunker, S.G. (1984) 'Modes of Extraction, Unequal Exchange, and the Progressive Underdevelopment of an Extreme Periphery: The Brazilian Amazon, 1600–1980,' *American Journal of Sociology*, 89(5) [Online]. Available at: https://www.journals.uchicago.edu/doi/abs/10.1086/227983 (Accessed: 20 March 2021)

Burger, M. and Gundlach, J. (2017) *The Status of Climate Change Litigation. A Global Review.* Available at: https://wedocs.unep.org/bitstream/handle/20.500.11822/20767/climate-change-litigation.pdf?sequence=1&isAllowed=y (Accessed: 20 March 2021)

Business Roundtable (2019) 'Business Roundtable Redefines the Purpose of a Corporation to Promote 'An Economy That Serves All Americans,' *Business Roundtable*, 19 August [Online]. Available at: https://www.businessroundtable.org/business-roundtable-redefines-the-purpose-of-a-corporation-to-promote-an-economy-that-serves-all-americans (Accessed: 20 March 2021)

Callen, T. (2020) 'Gross Domestic Product: An Economy's All,' *International Monetary Fund*, 24 February [Online]. Available at: https://www.imf.org/external/pubs/ft/fandd/basics/gdp.htm (Accessed: 20 March 2021)

Carroll, A.B. (2008) 'A History of Corporate Social Responsibility: Concepts and Practices' in Crane, A., McWilliams, A., Matten, D., Moon, J. and Siegel, D.S. (eds.) *The Oxford Handbook of Corporate Social Responsibility*. Oxford: Oxford University Press.

Catalyst (2021) *CEO Activism: Trend Brief*. Available at: https://www.catalyst.org/wp-content/uploads/2018/12/CEO-Activism_2.23.2021.pdf (Accessed: 20 March 2021)

Chanel (2020) *Chanel Announces Climate Commitments* [Press release], 10 March. Available at: https://www.chanel.com/us/climate-report/ (Accessed: 20 March 2021)

Chatterji, A.K. and Toffel, M.W. (2018) 'The New CEO Activists,' *HBR*, January–February [Online]. Available at: https://hbr.org/2018/01/the-new-ceo-activists (Accessed: 20 March 2021)

Chatterji, A.K. and Toffel, M.W. (2019) *Assessing the Impact of CEO Activism*. Available at: https://www.hbs.edu/ris/Publication%20Files/16-100._0fc0baf0-1735-47f2-a2c3-82e304e6e990.pdf (Accessed: 20 March 2021)

Clark, D. (2019) 'Delivering Shipment Zero, a Vision for Net Zero Carbon Shipments,' [Blog] *About Amazon*. Available at: https://www.aboutamazon.com/news/sustainability/delivering-shipment-zero-a-vision-for-net-zero-carbon-shipments (Accessed: 20 March 2021)

CO_2.Earth (2021) *CO_2.Earth Are We Stabilizing Yet?* [Online]. Available at: https://www.co2.earth/ (Accessed: 21 August 2017)

Cone Communications (2016) *2016 Cone Communications: Millennial Employee Engagement Study* [Press release]. 2 November. Available at: https://static1.squarespace.com/static/56b4a7472b8dde3df5b7013f/t/5819e8b303596e3016ca0d9c/1478092981243/2016±Cone±Communications±Millennial±Employee±Engagement±Study_Press±Release±and±Fact±Sheet.pdf (Accessed: 20 March 2021)

Country Economy (2020) '*Australia GDP – Gross Domestic Product* [Online],' *Country Economy*. Available at: https://countryeconomy.com/gdp/australia (Accessed: 20 March 2021)

Crippa, M. et al. (2019) *Fossil CO_2 and GHG Emissions of All World Countries*. Available at: https://op.europa.eu/en/publication-detail/-/publication/9d09ccd1-e0dd-11e9-9c4e-01aa75ed71a1/language-en (Accessed: 20 March 2021)

Desai, P. (2018) 'GDP is Destroying the Planet. Here's an Alternative,' *World Economic Forum*, 31 May [Online]. Available at: https://hbr.org/2011/01/the-big-idea-creating-shared-value (Accessed: 20 March 2021)

EEA, European Environment Agency (2020) *Total Greenhouse Gas Emission Trends and Projections in Europe* [Online]. EEA. Available at: https://www.eea.europa.eu/data-and-maps/indicators/greenhouse-gas-emission-trends-7/assessment (Accessed: 20 March 2021)

Elkington, J. (2004) 'Enter the Triple Bottom Line' in Richardson, J. and Henriques, A. (eds.) *The Triple Bottom Line: Does It All Add Up?* New York, NY: Routledge.

Elkington, J. (2020) *Green Swans: The Coming Boom in Regenerative Capitalism*. New York, NY: Fast Company Press.

EPA (2021) *Inventory of U.S. Greenhouse Gas Emissions and Sinks*. [Online]. EPA. Available at: https://www.epa.gov/ghgemissions/inventory-us-greenhouse-gas-emissions-and-sinks (Accessed: 20 March 2021)

European Commission (2019) *European Economic Forecast Spring 2019*. Available at: https://ec.europa.eu/info/sites/info/files/economy-finance/ip102_en.pdf (Accessed: 20 March 2021)

Fink, L. (2019) *Larry Fink's 2019 Letter to CEOs. Purpose & Profit* [Online]. BlackRock. Available at: https://www.blackrock.com/corporate/investor-relations/2019-larry-fink-ceo-letter (Accessed: 20 March 2021)

Fossil Free: Divestment (2019) *1200+ Divestment Commitments* [Online]. Fossil Free: Divestment. Available at: https://gofossilfree.org/divestment/commitments/ (Accessed: 20 March 2021)

FRED (2020) *FRED Real Gross Domestic Product* [Online]. FRED Economic Research. Available at: https://fred.stlouisfed.org/series/GDPCA (Accessed: 20 March 2021)

Friedman, M. (1970) 'The Social Responsibility of Business Is to Increase Its Profits,' *The New York Times Magazine*, 13 September [Online]. Available at: http://umich.edu/~thecore/doc/Friedman.pdf (Accessed: 20 March 2021)

Fullerton, J. (2015) *Regenerative Capitalism. How Universal Principles and Patterns Will Shape Our New Economy*. Available at: http://capitalinstitute.org/wp-content/uploads/2015/04/2015-Regenerative-Capitalism-4-20-15-final.pdf (Accessed: 20 March 2021)

Gelles, D. (2018) 'The C.E.O. Who Stood Up to President Trump: Ken Frazier Speaks Out,' *The New York Times*, 19 February, [Online]. Available at: https://www.nytimes.com/2018/02/19/business/merck-ceo-ken-frazier-trump.html (Accessed: 20 March 2021)

Gond, J.P. and Crane, A. (2010) 'Corporate Social Performance Disoriented: Saving the Lost Paradigm?,' *Business & Society*, 49(2) [online]. Available at: https://www.researchgate.net/publication/249705627_Corporate_Social_Performance_Disoriented_Saving_the_Lost_Paradigm (Accessed: 20 March 2021)

Harvey, D. (2007) 'Neo Liberalism as Creative Destruction: Annals of the American Academy of Political and Social Science,' *The Annals of the American Academy of Political and Social Science*, 610(1) [Online]. Available at: https://journals.sagepub.com/doi/abs/10.1177/0002716206296780 (Accessed: 20 March 2021)

Hausfather, Z. (2017) 'Analysis: Why US Carbon Emissions Have Fallen 14% Since 2005,' *CarbonBrief*, 15 August [Online]. Available at: https://www.carbonbrief.org/analysis-why-us-carbon-emissions-have-fallen-14-since-2005 (Accessed: 20 March 2021)

Hedblom, D., Hickman, B.R. and List, J.A. (2019) *Toward an Understanding of Corporate Social Responsibility: Theory and Field Experimental Evidence*. National Bureau of Economic Research Working Paper No 26222. Available at: https://www.nber.org/papers/w26222 (Accessed: 20 March 2021)

IEA (2019) *Global Energy & CO_2 Status Report 2019*. Available at: https://www.iea.org/reports/global-energy-co2-status-report-2019 (Accessed: 20 March 2021)

IEA (2020) *Global Energy Review 2020*. Available at: https://www.iea.org/reports/global-energy-review-2020/global-energy-and-co2-emissions-in-2020 (Accessed: 20 March 2021)

Jackson, T. (2017) *Prosperity without Growth: Foundations for the Economy of Tomorrow*. New York, NY: Routledge.

Jiang, R., Zhou, Y. and Li, R. (2018) 'Moving to a Low-Carbon Economy in China: Decoupling and Decomposition Analysis of Emission and Economy from a Sector Perspective,' *MDPI, Open Access Journal*, 10(4) [Online]. Available at: https://ideas.repec.org/a/gam/jsusta/v10y2018i4p978-d138240.html (Accessed: 20 March 2021)

Kapoor, A. and Debroy, B. (2019) 'GDP Is Not a Measure of Human Well-Being,' *HBR*, 4 October [Online]. Available at: https://hbr.org/2019/10/gdp-is-not-a-measure-of-human-well-being (Accessed: 20 March 2021)

Kering (2019) *Environmental Profit & Loss* [Online]. Kering Sustainability. Available at: https://www.kering.com/en/sustainability/environmental-profit-loss/ (Accessed: 20 March 2021)

Langley, M. (2016) 'Salesforce's Marc Benioff Has Kicked Off New Era of Corporate Social Activism,' *The Wall Street Journal*, 2 May [Online]. Available at: https://www.wsj.com/articles/salesforces-marc-benioff-has-kicked-off-new-era-of-corporate-social-activism-1462201172 (Accessed: 20 March 2021)

Lazonick, W., Mazzucato, M. and Tulum, Ö. (2013) 'Apple's Changing Business Model: What Should the World's Richest Company Do with all Those Profits?,' *Accounting Forum*, 37(4) [Online]. Available at: https://www.sciencedirect.com/science/article/abs/pii/S0155998213000434 (Accessed: 15 November 2020)

Lazonick, W. and Mazzucato, M. (2013) 'The Risk-Reward Nexus in the Innovation-Inequality Relationship: Who Takes the Risks? Who Gets the Rewards?,' *Industrial and Corporate Change*, 22(4) [Online]. Available at: https://ec.europa.eu/futurium/en/system/files/ged/ind_corp_change-2013-lazonick-1093-128_0.pdf (Accessed: 20 March 2021)

List, J.A. and Momemi, F. (2017) *When Corporate Social Responsibility Backfires: Theory and Evidence from a Natural Field Experiment*. National Bureau of Economic Research Working Paper No 24169. Available at: https://www.nber.org/papers/w24169 (Accessed: 20 March 2021)

Malm, A. (2018) *The Progress of This Storm: Nature and Society in a Warming World*. New York: Verso.

Margolis, J.D. and Walsh, J.P. (2003) 'Misery Loves Companies: Rethinking Social Initiatives by Business,' *Administrative Science Quarterly*, 48(2) [Online]. Available at: https://www.jstor.org/stable/3556659?origin=crossref&seq=1 (Accessed: 20 March 2021)

Mayer, J. (2016) *Dark Money: The Hidden History of the Billionaires Behind the Rise of the Radical Right*. New York, NY: Anchor Books.

Mazzucato, M. (2011) *The Entrepreneurial State: Debunking Public vs Private Sector Myths*. NY: Public Affairs.

Mazzucato, M. (2018a) 'Mission-Oriented Innovation Policies: Challenges and Opportunities,' *Industrial and Corporate Change*, 27(5) [Online]. Available at: https://academic.oup.com/icc/article/27/5/803/5127692 (Accessed: 20 March 2021)

Mazzucato, M. (2018b) *The Value of Everything: Making and Taking in the Global Economy*. London: Allen Lane.

Meier, S. and Cassar, L. (2018) 'Stop Talking About How CSR Helps Your Bottom Line,' *HBR*, 31 January [Online]. Available at: https://hbr.org/2018/01/stop-talking-about-how-csr-helps-your-bottom-line (Accessed: 20 March 2021)

Millman, O. (2020) 'Amazon Threatened to Fire Employees for Speaking Out on Climate, Workers Aay,' *The Guardian*, 2 January [Online]. Available at: https://www.theguardian.com/technology/2020/jan/02/amazon-threatened-fire-employees-speaking-out-climate-change-workers-say (Accessed: 20 March 2021)

Mitchell, T. (2011) *Carbon Democracy: Political Power in the Age of Oil*. London/Brooklyn: Verso.

Oberle, B. et al. (2019) *Global Resources Outlook 2019*. Available at: https://www.resourcepanel.org/reports/global-resources-outlook (Accessed: 20 March 2021)

OECD (2015) *Better Life Index* [Online]. OECD Better Life Index. Available at: http://www.oecdbetterlifeindex.org/#/11111111111 (Accessed: 20 March 2021)

Oreskes, N. and Conway, E.M. (2010) *Merchants of Doubt: How a Handful of Scientists Obscured the Truth on Issues from Tobacco Smoke to Global Warming*. London: Bloomsbury.

Orlitzky, M., Schmidt, F.L. and Rynes, S.L. (2003) 'Corporate Social and Financial Performance: A Meta-Analysis', *Organization Studies*, 24(3) [Online]. Available at: https://www.researchgate.net/publication/241180100_Corporate_Social_and_Financial_Performance_A_Meta-Analysis (Accessed: 20 March 2021)

Peakon (2020) *The Employee Expectations Report 2020*. Available at: https://peakon.com/heartbeat/reports/employee-expectations-2020/ (Accessed: 20 March 2021)

Porter, M.E. (1979) 'How Competitive Forces Shape Strategy,' *HBR*, March [Online]. Available at: https://hbr.org/1979/03/how-competitive-forces-shape-strategy (Accessed: 20 March 2021)

Porter, M.E. and Kramer, M.R. (2011) 'Creating Shared Value,' *HBR*, January–February [Online]. Available at: https://hbr.org/2011/01/the-big-idea-creating-shared-value (Accessed: 20 March 2021)

Povaddo (2017) *Corporate America's POV. A Povaddo Survey Examining Corporate Activism and Employee Engagement Inside Fortune 1000 Companies*. Available at: http://www.povaddo.com/downloads/Povaddo_Corporate_America's_POV_May_2017.pdf (Accessed: 20 March 2021)

Raftopoulos, M. (2017) 'Contemporary Debates on Social-Environmental Conflicts, Extractivism and Human Rights in Latin America', *The International Journal of Human Rights*, 21(4) [Online]. Available at: https://www.tandfonline.com/doi/pdf/10.1080/13642987.2017.1301035 (Accessed: 15 November 2016)

Setzer, J. and Byrnes, R. (2019) *Global Trends in Climate Change Litigation: 2019 Snapshot*. Available at: https://www.lse.ac.uk/GranthamInstitute/wp-content/uploads/2019/07/GRI_Global-trends-in-climate-change-litigation-2019-snapshot-2.pdf (Accessed: 20 March 2021)

Schaninger, B. et al. (2020) 'Demonstrating Corporate Purpose in the Time of Coronavirus,' *McKinsey & Company*, 31 March [Online]. Available at: https://www.mckinsey.com/business-functions/organization/our-insights/demonstrating-corporate-purpose-in-the-time-of-coronavirus (Accessed: 20 March 2021)

Sharma, N. Smeets, B. and Tryggestad, C. (2019) 'The Decoupling of GDP and Energy Growth: A CEO Guide,' *McKinsey Quarterly*, 24 April [Online]. Available at: https://www.mckinsey.com/industries/electric-power-and-natural-gas/our-insights/the-decoupling-of-gdp-and-energy-growth-a-ceo-guide# (Accessed: 20 March 2021)

Schwab, K. (2019) 'Davos Manifesto 2020: The Universal Purpose of a Company in the Fourth Industrial Revolution,' *World Economic Forum*, 2 December [Online]. Available at: https://www.weforum.org/agenda/2019/12/davos-manifesto-2020-the-universal-purpose-of-a-company-in-the-fourth-industrial-revolution/ (Accessed: 20 March 2021)

Smil, V. (2019) *Growth: From Microorganisms to Megacities*. Cambridge: The MIT Press.

Smith, R.E. (2011) *Defining Corporate Social Responsibility: A Systems Approach for Socially Responsible Capitalism*. MPhil Thesis. University of Pennsylvania. Available at: https://repository.upenn.edu/od_theses_mp/9/ (Accessed: 20 March 2021)

Stiglitz, J.E., Sen, A. and Fitoussi, J.P. (2010) *Mismeasuring Our Lives: Why GDP Doesn't Add Up*. New York, NY: The New Press.

Stiglitz, J.E. (2018) 'Beyond GDP,' *Project Syndicate*, 3 December [Online]. Available at: https://www.project-syndicate.org/commentary/new-metrics-of-wellbeing-not-just-gdp-by-joseph-e-stiglitz-2018-12?barrier=accesspaylog (Accessed: 20 March 2021)

Talberth, J. (2010) 'Measuring What Matters: GDP, Ecosystems and the Environment,' [Blog] *World Resources Institute*. Available at: https://www.wri.org/blog/2010/04/measuring-what-matters-gdp-ecosystems-and-environment (Accessed: 20 March 2021)

The B Team (2017) 'CEOs of Major U.S. Companies Urge Trump: Stay in Paris,' *The B Team*, 10 May [Online]. Available at: https://bteam.org/our-thinking/news/30-major-ceos-call-on-trump-stay-in-paris (Accessed: 20 March 2021)

The British Academy (2019) *Principles for Purposeful Business*. Available at: https://www.thebritishacademy.ac.uk/documents/224/future-of-the-corporation-principles-purposeful-business.pdf (Accessed: 20 March 2021)

The Shift Project (2012) 'Is the Green GDP a flawed index?,' *The Shift Project*, 31 July [Online]. Available at: https://theshiftproject.org/en/article/is-the-green-gdp-a-flawed-index/ (Accessed: 20 March 2021)

UNEP, United Nations Environment Programme (2019) *Emissions Gap Report 2019*. Available at: https://www.unep.org/resources/emissions-gap-report-2019 (Accessed: 20 March 2021)

Vasilogrambros, M. (2014) 'Hobby Lobby Ally Invokes the 'Chick-fil-A' Defense,' *The Atlantic*, 30 June [Online]. Available at: https://www.theatlantic.com/politics/archive/2014/06/hobby-lobby-ally-invokes-the-chick-fil-a-defense/437958/ (Accessed: 20 March 2021)

We Mean Business (2015) *The Business Brief. Shaping A Catalytic Paris Agreement*. Available at: https://s3.amazonaws.com/assets.wemeanbusinesscoalition.org/wp-content/uploads/2015/11/03183842/The-Business-Brief.pdf (Accessed: 20 March 2021)

Wells, W. (2012) '"Corporation Law is Dead": Heroic Managerialism, the Cold War, and the Puzzle of Corporation Law at the Height of the American Century,' *University of Pennsylvania Journal of Business Law*, 15 [online]. Available at: https://www.law.upenn.edu/live/files/1807-wells15upajbusl3052013pdf (Accessed: 15 November 2016)

White, B. and Meese, D. (2020) 'How Purpose Lights the Way During Covid-19,' *BrightHouse*, 6 May [Online]. Available at: https://www.thinkbrighthouse.com/2020/05/how-purpose-lights-the-way-in-crisis/ (Accessed: 20 March 2021)

WMO (2019) *WMO Statement on the State of the Global Climate in 2019*. Available at: https://library.wmo.int/doc_num.php?explnum_id=10211 (Accessed: 20 March 2021)

Wu, Y. et al. (2019) 'Decoupling China's Economic Growth from Carbon Emissions: Empirical Studies from 30 Chinese Provinces (2001–2015),' *Science of The Total Environment*, 656 [Online]. Available at: https://www.sciencedirect.com/science/article/abs/pii/S0048969718347405 (Accessed: 20 March 2021)

3
THE GREATEST ECONOMIC OPPORTUNITY IN HISTORY
Building a new climate economy

Climate action is one of the greatest economic opportunities in history – an opportunity to increase investments and profits and create new markets but also an opportunity to transform the economic system in a way that is more sustainable and aligned with the requirements of science. Transformative cycles in the economy are relatively short: of four of the five largest companies in the world by market capitalization, two didn't exist 45 years ago, Microsoft and Apple, and the other two, Amazon and Alphabet, did not exist 25 years ago at the time of writing. Will the new climate economy yield equally large gains in a short span? Will it lead to the development of new industries altogether? Will it transform the economy for the next quarter-century or for centuries to come? How will this transformation – whichever shape it will take – occur? This chapter will not be about predictions but about tracking opportunities and signals of change in the current economy.

Recent research by The New Climate Economy has shown that climate action could deliver a net economic gain of US$26 trillion globally and cumulatively until 2030 (about one-fifth of the current global economy), compared to business as usual (Mountford et al., 2018, p.8). For comparison, it is expected that in that time period, total investment in new infrastructure alone will amount to US$90 trillion, which is more than the current stock of infrastructure (The New Climate Economy, 2016, p.2). The scale of investment and the scale of economic gain signal very clearly that the economy is at the dawn of, and possibly even in the very early stages, of a radical transformation.

The view that climate action is not consistent with other business goals, or that it is contrary to them, is not supported by evidence; rather, evidence points to it being an opportunity for all economic actors in various dimensions. Signs that the economic opportunity of climate action, defined by shifts in dollar flows, are visible. This evidence chips away at the notion that climate action is "too costly" and that it is in opposition to the goals of rapid economic growth - and in developing countries, that it would slow down, in the best of cases, development and poverty reduction. In this chapter, we will explore more evidence that climate action represents a genuine economic opportunity.

However, the opportunity is not only an economic one. Mobilizing a company and its resources to develop and implement a climate strategy requires understanding the internal logic of companies - as parts of a system but also as systems unto themselves. Decision-making in these systems is governed by different imperatives buttressed by structures and processes that also reflect them. In addition to structures of authority - or "chains of command" - decisions in companies are directed by a number of business goals.

The following taxonomy of goals is important since it helps us highlight a distinction that is often used to question the reasons a company would have to raise its climate ambition. Climate action is often characterized as something that companies do above and beyond their business priorities. It is not seen to be in company's narrow self-interest - "there's nothing in it for them." As a result, climate action is often within the remit of corporate social responsibility (CSR) initiatives or even corporate foundation programs, far from the core business of a company. As companies do not see their so-called self-interest in investing in climate action, they do not integrate it to their business and prefer pilot programs or programs that are targeted to very specific arms of the business. These initiatives tend to be voluntary, limited in time or very short-term, on a small scale relative to the entire business, and rarely designed to generate revenue, and most importantly, divorced from the core business.

Climate leadership, however, does not consist of these sorts of programs. The first step in understanding what it is requires understanding how companies organize decision-making. The view that climate action and corporate self-interest are opposed to each other can be weakened if not entirely debunked by considering the four types of business goals presented here. Self-interest - or the interest of a company in continuing to exist and grow - is a driver for business decisions. And hence in order to build the argument that climate action and climate leadership are inevitably charting the path of the future economy, it is necessary to start from the vantage of the self-interest of a company. There may be other reasons for a company to raise its ambition on climate, and we will explore them in later chapters - corporate citizenship, purpose, and so on. But without a clear picture of the drivers of corporate decision-making, it is impossible to activate the levers necessary to raise climate ambition and to generate climate leadership.

When we think of the self-interest of a company, we often think of financial and economic goals. Hence there is a tendency to associate the opportunity of climate action in dollar terms only. In what follows, we will see that on all four dimensions of business goals below, the current moment for raising climate ambition is historical. Though different companies, in different sectors, in different markets, of various sizes, led and manned by different people do not form

TABLE 3.1 Higher-order business goals

Economic goals	These goals are often taken to be the primary objectives of a business. Under this umbrella, terms such as "profit," "revenue," "growth" bring multiple dimensions to a single decision-making factor: the focus on securing or growing the financial assets of a company.
Strategic goals	Strategic goals are those that allow a company to pursue its competitive advantage in a given market. There are many definitions of strategy and they all share the notion of strategy being the bridge between high-order goals on the one hand and tactics or concrete actions on the other, with those goals having to do with the competitive advantage and continued existence or growth of the company.
Environmental goals	These goals are rarely amongst the most prominent ones. Here I am distinguishing them from the kinds of environmental goals that fall under the remit of corporate social responsibility (CSR) but understanding them rather as whatever goals help a company maintain the kinds of relationships to the environment that will allow it to continue to produce its goods or deliver its services. This assumes that for many companies, natural resources or natural environments will be essential to the company's activities. These can hence extend to managing supply chains that secure access to these resources and environments. These goals will vary depending on the company and on its sector; indeed, some companies will have competing if not contrary goals compared to others. For instance, it could be the integrity of a waterway for transport; the availability of timber as raw material for transformation into paper; or the reliability of snowfall for a ski resort.
Social goals	Similar to environmental goals, social goals here are not understood as the typical CSR objectives but rather goals related to the social environments within which companies operate and that require an amount of stability, levels of skills, the development and availability of technological advancements, or availability of labor itself. Sometimes achieving these goals is akin to seeking a *license to operate*.

a monolithic block, it is possible to identify four types of higher order business goals that apply across the board (see Table 3.1).

The opportunity represented by climate action can be described along these four dimensions of corporate decision making, and together they demonstrate that this is a moment with the potential to transform not only how companies operate but how entire industries function. These dimensions of business decision-making will be treated separately for purposes of clarity in presentation; however, they are not conceptually independent and companies' decisions may be determined by a combination of all of these goals. The remainder of this chapter will show how these business goals interplay with climate action to yield transformative opportunities in key areas.

Before delving into a more detailed analysis, context about the economy as a whole is helpful to show that we are already observing a global trend toward investment in climate action. While the trend is only starting to show signs of momentum, it is also moving the economy wholesale in the same direction.

Where companies in the private sector are concerned, action is already underway, but the multi-trillion-dollar opportunity – and the associated emissions reduction and resilience building – remains on the table. Some changes have already steered the economy toward embedding climate action into business decision-making. Companies and financial institutions are already encouraged or incentivized to disclose key carbon, environmental, social, and governance data and to integrate them into risk analysis, business models, and investment decision-making. Examples of notable shifts in the economy abound:

- The Financial Stability Board Task Force on Climate-related Financial Disclosures (TCFD) established a market-driven initiative in 2016 with guidelines for voluntary and consistent climate-related financial risk disclosures in mainstream financial filings. More than 160 financial firms responsible for over US$86 trillion in assets have committed to support the recommendations of the TCFD (Mountford et al., 2018, p.13). In addition, on average across the TCFD recommendations, 42% of companies with a market capitalization greater than $10 billion disclosed at least some information in line with each individual TCFD recommendation in 2019 (The Financial Stability Board Task Force on Climate-Related Financial Disclosures, 2020, p.8).
- Green bond issuance in 2019 exceeded that of previous years to reach US$250 billion, with more green bonds hitting the market in the first six months on the year than in 2018 altogether (Climate Bonds Initiative, 2021). In 2020, that number rose to US$270 billion (Jones, 2021).
- BlackRock Inc.'s climate pledge has shown the extent and nature of the financial opportunity it represents shortly after it was rolled out. An exchange-traded fund from Blackrock has attracted more than $600 million in one week (the week of February 12, 2020), according to data compiled by Bloomberg. That's the best debut for any US ETF in 2020, and a sign BlackRock isn't the only one that sees climate change as a defining factor in companies' long-term prospects, but rather that the opportunity for money managers is real and ready (Ballentine and Gupta, 2020; Massa, 2020).
- Climate Action 100+, launched in 2017, is another investor-led initiative backed by more than 360 investors with US$34 trillion in assets under management with a goal to influence targeted high-emitting companies to implement strong governance frameworks, take action to reduce greenhouse gas emissions across the value chain consistent with the Paris Agreement, and to provide enhanced corporate disclosure in line with the TCFD.
- The New Climate Economy estimated in 2015 that globally the private sector accounts for a US$5.5 trillion market in low-carbon and other environmental technologies, products, and services. This figure is likely to be higher now due to a shift in cost structure for renewables (Jacobs and Mountford, 2015, p.10).

These signals of change point toward the direction of a transformation of the economy, especially to the extent that it is measured in financial terms. This lens is a good indicator of change, comparing like to like, that is, dollars to dollars. The opportunity for business is, indeed, a financial opportunity, but it is not only a financial one. As we have already seen, there are other business goals that are served by the transition to a carbon-neutral economy. A genuine opportunity for business will indeed serve a combination of goals, not only financial ones. There is value in using tried and true indicators of opportunity – such as growth, GDP, Return on Investment, Assets under Management, and so on. This book, however, rests on the premise that old frameworks will not be adapted to the new economy that is profiling itself on the horizon. The new economy will require new frameworks and new indicators because it will be rooted in; indeed, it will yield, a new reality. It is even likely that our current models will underestimate the benefits of the new climate economy and that our current economic models are inadequate to capture opportunities fully. For now, however, let us turn to how the combination of business goals and financial opportunity plays out in different key sectors that have a ripple effect on the entire economy.

Energy systems

Electricity generation and heat production are the largest sources of CO_2 emissions, according to the IPCC AR5 report, accounting for a quarter of global emissions (IPCC, 2014, p.123). Progress in these areas represents a key lever for mitigation and climate action – and an equally important opportunity for companies. All sectors of the economy will be affected by the transformation of energy systems.

In all areas of the energy system, the impacts of the COVID-19 pandemic have been felt: through blows to global trade, a slowing pace of consumption, investments, and ensuing technological developments. Nonetheless, the trends described below, if decelerating, still hold: the job market in new energy systems has remained resilient, some technological advances have occurred, and implementation of solutions has continued (Turk and Kamiya, 2021).

Electricity generation

Context

Energy systems are already undergoing a deep transformation at pace. At the time of writing, 102 cities procure at least 70% of their electricity from renewable sources (UNFCCC, 2018). This trend is representative of a transition in electricity procurement from companies too and is creating advances in innovation, technological advances, and market competitiveness.

Global solar capacity grew faster than all fossil fuels combined, including coal-, oil-, and gas-fired power stations, for the first time in 2017, with renewables accounting for almost two-thirds of net new power capacity the previous year in 2016. In addition, renewables become cheaper than grid electricity in key markets including China and the United States (EY, 2019), as renewable energy grows into an increasingly unsubsidized power source (Brown, 2019).

Investments in research and the development of new technologies, as well as smart policy frameworks that favor the integration of renewables into grids and energy systems, has led to renewables being cheaper at time of writing than from fossil fuels in most regions of the world. IRENA shows that renewables are the lowest-cost source of new power generation in most parts of the world as of 2018 (Anuta, Ralon and Taylor, 2018, p.9). Indeed, on a global level, the cost of renewable energy continues to fall. Certain technologies such as onshore wind and utility-scale solar became cost-competitive with conventional generation in the last decade on a new-build basis. It is now competitive with the marginal cost of existing conventional generation technologies (Lazard, 2019). Among projects due to be commissioned in 2020 in the IRENA and PPA database, 77% of the onshore wind and 83% of the utility-scale solar PV project capacity will produce electricity cheaper than fossil fuel-fired power generation for new generation (Anuta, Ralon and Taylor, 2018, p.9).

As of 2019, approximately 74% of coal generation in the United States could be replaced by wind and solar with direct savings to end-use customers. By 2025, this number could reach 86% (Gimon et al., 2019, p.1). In addition, and for the first time, the rapidly falling costs of batteries are allowing optimized combinations of renewable sources of electricity to be cost-competitive with gas-fired generation without jeopardizing the reliability of services (RMI, 2019, p.7).

In the European Union, it is feasible that the share of renewable in the energy mix could double by 2030 – with estimated savings of up to $25 billion per year by 2030, with similar cost-effectiveness (IRENA and European Commission, 2018, p.20).

China has become a world leader expanding from a small, local, and mostly rural solar industry in the 1990s to the largest in the world today – not only in terms of solar installations but also in terms of electricity generation (Edmond, 2019). In 2018, China accounted for more than a third of the worldwide total solar installed capacity and for half of new solar installations (IRENA, 2018), allowing a fifth of cities with solar installations to produce electricity cost-competitively with coal (Yan, 2019, p.709).

In Africa, solar panel prices fell 80% between 2009 and 2016 and are able to meet energy demand both on- and off-grid, a key benefit for the continent and its provision of access to energy. The infrastructure is diverse in this sense and has led to the emergence of solar home systems that are able to provide electricity needs for a year for off-grid households for much less than the average price for sub-optimal on-grid energy services (Taylor and Young So, 2016, p.7).

Opportunity related to economic goals

These figures in themselves suggest that an opportunity to drive down costs of renewables is possible. Given that the private sector, via procurement and investment, played a central role in driving renewable energy deployment, we can expect more gains to be reaped from continuing such strategies. Indeed, corporate sourcing of renewables more than doubled during 2018. Global investment in renewables did decrease compared to 2017, but developing and emerging economies provided over half of all investment in 2018 – suggesting that the greatest opportunity lies there.

To give a sense of the scale of renewables procurement, RE100, a global corporate leadership initiative bringing together influential businesses committed to 100% renewable electricity, has a membership of global multinationals that would be the 21st largest electricity consumer in the world. The collective electricity consumption of this group is on par with a medium-sized country, with a 228 TWh/year consumption in 2018, comparable to Indonesia or South Africa. That's an extra 40 TWh/year in demand for clean energy globally.

Companies are already availing of the economic opportunity of renewables. T Mobile USA, for instance, projects US$100 million in savings between 2018 and 2033 (RE100, 2018, p.8).

Opportunity related to strategic goals

As renewable electricity networks become more advanced, procurement becomes a pathway to innovation and competitive advantage. Specifically, by innovating in their sourcing strategies, companies can create new and rapidly evolving markets. Internal factors such as capital availability and external factors such as market maturity result in mixed approaches that advance renewable energy deployment globally but that have different direct impacts on the grid.

According to RE100, the breakdown of sourcing strategies in 2017 was similar to 2016, with unbundled energy attribute certificates (EACs) and contracts with suppliers representing 81% of the total (RE100, 2018, p.10). PPAs also increased and represented 16% of renewable electricity consumed by RE100 companies, with the amount sourced this way reaching just under 9 TWh in 2017 – almost double the amount reported for 2016.

Opportunity related to environmental goals

Recall that environmental goals are goals related to securing access to the resources necessary to operate. Hence, supply chain management is key to achieving them. As a result, companies that shift their procurement to renewables encourage a similar shift across their supply chains. RE100 member Apple, for instance, has bolstered its Supplier Clean Energy Program with a new $300 million investment fund in China to support more of its suppliers to switch to clean power.

Opportunity related to social goals

Where social goals are concerned, companies can display leadership in corporate citizenship in three ways: by contributing to a just transition of labor toward this sector, by engaging in the sphere of policy, and by affecting consumer choices, and by reducing their negative impact on public health via emissions-related air pollution. In a world that has experienced a global pandemic for over 12 months at the time of writing, considerations about the resilience of the job market and public health are paramount for any actor with the power to affect them.

IRENA tells us that the energy sector employed 11 million people worldwide in 2018, which is 700,000 more than the previous year – and that this trend is on the rise. Given that the shift from fossil fuels will lead to a transformation of labor employed in energy, providing these jobs and the trainings required to develop a competent and skilled labor force is an opportunity for companies to avert a crisis in employment (IRENA, 2019, p.3).

One million dollars spent on fossil fuels in the U.S. will generate 2–3 full-time equivalent jobs, while $1 million spent on renewable energy or energy efficiency will create 7–8 jobs, according to a 2017 study (Garrett-Peltier, 2017). These jobs tend to provide higher and more equitable wages when compared to all workers in the United States (Muro et al., 2019).

Companies that have adopted a renewable electricity procurement strategy have the opportunity to advocate for favorable policies and policy frameworks that can unlock cost savings and reliable supply of renewables. Examples of successful contributions to the public conversation include RE100 members Mars, Unilever, and Fujitsu supporting RE100 at the Federal Parliament in Australia by demonstrating the potential of corporate sourcing of renewables to drive a prosperous economy.

As consumer and citizen awareness is rising, B2C companies and their brands can increase their impact and customer engagement by providing – and communicating about – products that are minimally based on fossil fuels, or not at all. The world's largest brewer, AB InBev, has done just that by developing a 100% renewable label in 2018 in the US.

Renewables installations can also generate revenue as taxes and lease payments to farmers and landowners who host them on their land. Wind turbine hosting in the United States generated US$289 million in lease payments in 2018 (EIA, 2021b), and wind farms paid $761 million in the same year in state and local taxes. In the context of declining farming incomes, these figures are essential to local rural economies.

Heat generation

Context

Renewables have seen far less growth in the heating and cooling sectors, with progress mostly due to a lack of strong policy support and by slow developments in new technologies both on the supply side (for implementation) and on the demand side (for use) (Chassein, Roser and John, 2017, p.8).

Global demand for thermal energy, including the end-uses of heating and cooling, accounts for around half of final energy consumption (IRENA, IEA and REN21, 2018, p.93). Of these, demand for heat is the largest share, thought demand for energy for cooling is growing too for private use by individuals and even faster in large-scale projects and heavy consumers such as mining (IEA, 2018, pp.10, 290; IRENA, IEA and REN21, 2018, p.25). According to the IEA, energy for heating and cooling is largely fossil fuel–based and contributed almost 40% of global energy-related CO_2 emissions (IRENA, IEA and REN21, 2018, p.25).

The sectors most concerned by heat generation are buildings and industry, which both consume similar amounts of final thermal energy but with different shares of renewables involved. Where the built environment sector is concerned, renewables contribute about 9% of final heat consumption, mostly from bioenergy, solar thermal, or geothermal (IEA, 2018, p.4). For industry, that contribution is approximately 11%, the majority of which is also below 100°C (IEA, 2018, p.13). For temperatures under 200°C, renewables can supply heat, suggesting that further integration of renewable energy can increase as technologies develop and cost continues to decrease (BNEF, 2019, p.9).

Barriers to the advance of renewable energy in heating and cooling are technological as well as policy-related. Heat and cooling supply are highly localized and tend to be produced at the point of demand. The diversity of these points translates into diversity of markets, of technologies, and of cost structures. It also implies that there is not consolidated global industry in heating. Heating solutions powered by renewables will be highly tailored to the local technological and policy context. The latter is especially relevant with regards to the reliability of networks, how amenable they are to renewables, and favorable investment landscapes. Where policy is concerned, lack of ambition characterizes most jurisdictions, even as awareness of the urgency to address this is increasing (Decarb Heat, 2020; Heat Roadmap, 2020).

Opportunity related to economic goals

The opportunity for business is in the first instance related to cost savings as renewables continue to become cost competitive. Given the high localization of heating and cooling, these savings will be calculated on a localized basis, too, depending on variables such as the price of energy and the cost of connecting to new networks.

Opportunity related to strategic goals

As technology develops, local procurement of energy in markets where heating and cooling demand is increasing – that is, in emerging and developing economies as temperatures are rising – increases the resilience of business in the face of shifting energy markets.

Opportunity related to social goals

Given the relative lack of policy attention, opportunities exist for companies to advance strategic benefits via policy leadership and agenda-setting on these issues, most likely at local levels too.

Energy efficiency

Increasing energy efficiency – accomplishing the same output with less energy use – is a key lever to reducing emissions and is often construed as a low-hanging fruit given the low costs in investments relative to the benefits it offers. Measures around energy efficiency are among the most common and most ambitious types of measures.

Opportunity related to economic goals

In IEA member countries, advances in energy efficiency since 2000 have avoided around 20% more energy use in 2018, an amount greater than the final energy consumption of India that year. These savings in energy use resulted in the avoidance of over 15% or USD 600 billion more energy costs, with industry and services sector accounting for over half of these (IEA, 2020).

In the United States, for instance, efforts led by the Department of Energy aim to double energy efficiency by 2030 (Energy2030, 2015). This program estimates that this represents US$327 million savings per year in energy costs.

Cost savings tend to be the strongest driver for companies to prioritize investments in energy efficiency, and other drivers include greater energy security and futureproofing of operations. South African listed retailer, Woolworths Holdings Ltd, is a member of the EP100 initiative, launched in 2016 as a partnership between the Alliance to Save Energy and The Climate Group and bringing together a growing group of leading companies committed to smarter energy use. The company expects to save approximately US$10 million between 2019 and 2024 from using closed-door refrigeration across its stores – which is just one of its efficiency measures.

Opportunity related to strategic goals

An energy plan that focuses on energy efficiency supports strategic objectives by liberating resources for other investments, by increasing total factor productivity, and by favoring economic growth.

Furthermore, there is still scope for innovation in energy efficiency technologies, with opportunities to lead the sector and markets in this field.

Opportunity related to environmental goals

In terms of securing supply of energy, efficiency measures, by definition, improve supply chain management by reducing the quantity of electricity that is needed to operate at the same levels.

Opportunity related to social goals

Energy efficiency measures have added 1.3 million jobs to the US economy. In 2018, energy efficiency accounted for half of the energy industry's overall net new jobs and employed twice as many US workers as the entire fossil fuel industry, for a total of a little over 2.3 million (for comparison, the total number of wait staff in bars and restaurants in the United States was 2.25 million people) (E2, E4 The Future, 2019, p.2; EESI, 2019). According to the same report, these represent long-term opportunities for not only good jobs but careers that can't be outsourced. These include making new and existing buildings more efficient, upgrading, repairing heating, ventilation, and air conditioning (HVAC) equipment, manufacturing efficient and certified appliances, and designing and building high-performance constructions.

End users of energy reap large benefits from energy efficiency. Consumers benefit from energy-efficiency-related cost savings, with the average American family, for instance, saving US$500 a year on utility bills due to federal efficiency standards for appliances, lighting, and plumbing – with the benefits outweighing the costs by 5 to 1 (DeLaski and Mauer, 2017, p.iii).

Transport

The transport sector represents 23% of global greenhouse gas emissions globally – over half of which is from light vehicles – and is the fastest-growing contributor to climate change (Teter, 2020). In the United States alone, transport accounts for 29% of total emissions of 6.4 billion metric tons of greenhouse gas emissions (EPA, 2020).

Historically, emissions from transport have always been comparatively high. The US Energy Information Agency (EIA) started keeping records in the United States in 1973 and recorded more than 2 billion metric tons of greenhouse gas emissions from transportation for the first time in 2007. The 2012 recession saw those emissions an inch below the 2 billion mark, but they have been rising ever since, with, in 2018, transport emissions totaling 1.9 billion metric tons, just below the 2007 peak (EIA, 2021a).

Projections also place transport as a key sector in the fight against climate change and emissions. The Intergovernmental Panel on Climate Change (IPCC)'s AR5 report estimated that by 2050, the transport sector's emissions of greenhouse gases could reach 12 billion tons annually.

In addition to impact on climate, emissions from transport represent a public health crisis. Across the world, nine out of ten people breathe air containing high levels of pollutants, including from transport, according to the World Health Organization. About seven million premature deaths are due to the effects of air pollution every year. In addition to premature fatality, air pollution has a negative impact on many aspects of life: respiratory illnesses, days of missed work, lifelong reduced lung capacity for children who are exposed to it in childhood (WHO, 2019).

Principal measures to curb emissions include electrification of transport as well as innovations in transport modalities. Both of these offer opportunities for companies along all four dimensions of business goals.

Interestingly, the halt of most airborne transport on a global scale and of much transport in countries cities that have known lockdowns during the 2020 pandemic of COVID-19 has shown the direct and relatively immediate impacts of reducing our global dependency on fossil-fuel-based transport (Le Quéré, 2020, pp.647–653). Emission rates returned to pre-pandemic levels as soon as lockdown restrictions eased – showing that the main lesson to draw was the confirmation between fossil-fuel-based transport and air pollution and GHG emissions, which we had already drawn from decades of scientific studies (EEA, 2019; Krupnick, Rowe and Lang, 1997).

Opportunity related to economic goals

As companies shift their production from internal combustion engine vehicles to electric ones, so too are they creating new markets. Analysis projects that the global lithium-ion battery market will reach US$105 billion by 2025.

By electrifying their fleet, companies are also setting themselves up for cost savings. In the United States, for instance, gasoline costs US$2.25 on average at the time of writing, while charging an electric vehicle for the equivalent amount of energy costs only US$1.15 (Borlaug et al., 2020). Furthermore, the International Council of Clean Transportation estimates that electric vehicles will be cost-competitive on an upfront basis with internal combustion engine vehicles by 2024–2028 in the United States (Lutsey and Nicholas, 2019, p.11). For companies running buses, cost savings are also projected there: the fuel and maintenance costs for an electric bus over its lifetime are about US$400,000 less than a conventional bus (Casale and Mahoney, 2019, p.3).

LeasePlan, a member of the EV100 initiative that supports companies toward transitioning their fleets to full electrification, released its total cost of ownership (TCO) analysis for EVs in Europe in 2019, showing that across close to one thousand scenarios, EVs are cost-competitive to their conventional counterparts (LeasePlan, 2019, pp.3–11). The TCO of EVs turned out to be on average 5% lower than the diesel vehicles, without taking into considerations costs associated with strict restrictions on diesel emissions and internal combustion engine vehicles that are increasing in cities worldwide.

In 2019, the International Council for Clean Transportation published similar results showing that zero-emission trucks are expected to reach TCO parity with diesel trucks within the decade in the United States (Hall and Lutsey, 2019, p.1).

There are, however, few EVs on the road at the time of writing. In the United States, fewer than 2% of cars are electric, with similar figures in Europe and lower ones elsewhere (Joselow, 2018). Where rates are high, such as in Norway, where EVs account for about a third of sales, it is due to government incentives to boost sales (Knudsen and Doyle, 2019). The rate of increase in the US and in Europe is significant, with figures close to doubling in 2019; however, more must be done to fully transform the sector in line with upcoming policy deadlines in various regions. This includes driving sales, of course, but also policy engagement at the local level, investment in infrastructure, and R&D investment into batteries.

Opportunity related to strategic goals

Since electrification of the transport sector is key to its decarbonization, it is no surprise that there is immense momentum to end the internal combustion engine in favor of electric vehicles. Restrictive emissions and fuel-efficiency regulations are becoming more commons worldwide and are leading carmakers to roll out vehicles more able to fit within those restrictions. But companies are not only responding to policy – they are availing of the opportunity to lead on innovation, capturing new markets, and improving supply chain management. Most auto manufacturers have engaged in transforming their offer, capturing these benefits.

The 2020 pandemic has caused many auto manufacturers to shut plants temporarily, slowing down the production of EVs. However, some companies have leveraged the opportunity of reopening of focusing their production on EV, with the EV market much more likely to experience a quick recovery and strong growth globally (McKinsey& Company, 2020, p.6).

BMW Group aims to have electrified vehicles account for between 15% and 25% of sales by 2025 (BMWGroup, 2017). The company also expects electrified vehicle sales to increase twofold by 2021, as compared with 2019, and a 30% growth in those sales year over year through 2025.

Daimler has invested upwards of US$11 billion to electrify its fleet with a view to producing more than 10 EVs by 2022 (Daimler, 2017). The company plans to offer an electric option of every Mercedes-Benz model.

Fiat Chrysler will invest US$10.5 billion to electrify its offering through 2022 (Barry, 2018), at which point it projects it will offer at least 12 hybrid or all electric cars. This investment includes a focus on supply chain standards in order to ensure that new developments are ethically sound (Markets Insider, 2019).

Ford focused its investment in EVs in 2018 with a view to a 2022 model lineup with 40% EVs (Carey and White, 2018). One of its most recent electric models is the Mustang Mach-E, which resembles its iconic sports car (DeBord, 2019).

General Motors has positioned Cadillac to be the group's leading brand for EVs, with a plan to have at least more than half – and possibly all – of its models electric by 2030 (White, 2019). Manufacturing for an electric SUV, one of the first of its kind, is expected to start in 2023.

Honda has announced that every model sold in Europe by 2020 will be at least partially electrified, which is different from being fully electric (Szymkowski, 2019). This is scaling back from the company's previously stated ambition to release a full lineup of electric cars by 2025 (Honda, 2019).

Nissan plans to release eight new electric cars by 2022 (Zhang, 2018).

Toyota is to date the global leader in the hybrid vehicle market with upwards of 80% of the market (Buckland and Tajitsu, 2019), aims to generate half of its revenue from EVs by 2025.

Volkswagen plans to invest over US$30 billion to develop EVs until 2023 (Matousek, 2019) with a view to making them account for 40% of the cars they manufacture by 2030. The company estimates it will reach its target of 1 million electric cars on the road two years ahead of its forecast in 2023 (Volkswagen, 2019).

Volvo has unveiled its first EV, the XC40 Recharge, on sale in late 2020 (Hawkins, 2019). The brand has committed to ensuring that 50% of sales will be generated from EVs by 2025 and to reducing the carbon footprint of every vehicle it manufactures by 40% (Volvo, 2019b). As part of its plan to become climate neutral by 2040, the company plans to release a new EV every year until 2025 (Volvo, 2019a).

Opportunity related to environmental goals

The development of the battery market opens companies up to supply chain risks. Traceability and mapping of notoriously contentiously sourced raw materials – cobalt, lithium, or nickel – are essential to mitigate unethical practices that threaten the future for the communities where the raw materials are sourced. Herein lies an opportunity for companies to innovate and to design supply chains that meet these standards.

The Responsible Blockchain Sourcing Network is an industry collaboration using blockchain technology to improve traceability in support of responsible sourcing of mined raw material. With investment from Fiat Chrysler, Volvo, Volkswagen, as well as other key players in the supply chain, it aims to launch a program focused on cobalt in 2020 (Markets Insider, 2019).

Opportunity related to social goals

One of the main opportunities in this category consists in reducing emissions in highly congested urban areas so as to improve air quality and decrease negative impacts such as premature deaths or respiratory illnesses. Around 84% of the members of the Climate Group initiative EV100 cite this as the second most important driver for switching to fully electric fleets; indeed, four healthcare companies – AstraZeneca, Genentech, Mawdsleys, and Novo – joined the program in 2019 with health benefits from reduced air pollution as a key area of impact.

The great momentum displayed by auto manufacturers in transforming their industry is a key lever with policymakers and local, national, and international levels first, to match the private sector's ambition, and second to accelerate the market transformation already engaged. The opportunity for leadership on this is significant and is certain to have rapid and deep impact on the sector.

Land use and agriculture

Land use and agriculture, from which we derive food, fiber, timber, energy, and other ecosystem function, accounts for approximately 23% of total GHG emissions (IPCC, 2020, p.10). Land use is at the center of human livelihoods and societies, with human activities affecting more than 70% of global land surface (excluding ice), according to the IPCC (2020, p.7). Given the vastness of this area of activity, we cannot talk about agriculture as a homogenous field. What we do know is that as population is growing, so does our reliance on land and land-related activities that result in increasing net GHG emissions, loss of natural ecosystems (e.g., forests, savannahs, natural grasslands, and wetlands), and declining biodiversity, with regional variability (IPCC, 2020, p.7). Food and resource waste is also an increasing issue, with, according to the IPCC, between a quarter and a third of all food produced being lost or wasted and hence contributing to GHG emissions (2020, p.7).

Land use and agricultural practices are not only central to global emission, but they also represent a significant lever to reduce emissions as well as to capture them via carbon sinks. Sustainable land management is key to reducing the negative impacts of climate change on ecosystems and societies. Consequently, companies in the sector have an opportunity not only to reduce GHG emissions via mitigation measures but also by transforming the sector so as to increase its potential to restore ecosystems and reduce impacts of climate change.

Addressing climate change in this sector cannot be achieved piecemeal and must include a structural transformation of key systems from agricultural systems to consumption and including finance innovation as well as cultural shifts. There are no silver bullets in this transformation and no universal solution will fit each region and each culture. Hence it is likely that the movement to transform the agriculture and land use system will include a significant diversity of actors engaging in a diversity of types of actions that will necessarily result in co-benefits around the sustainability of livelihoods, human health, food security, and inclusive development.

Measures to address climate change through agriculture and land use typically fall under four categories that are valid for all types of actors including companies in the private sector: (i) improving diets, (ii) developing nature-based solutions, (iii) influencing consumption by transforming value propositions and value chains, (iv) ensuring inclusive development for all actors.

Specifically, for business, this means that a series of actions are available, including, for instance:

1. Improving diets
   a. Shift R&D and marketing resources into healthier food options
2. Developing nature-based solutions
   a. Establish science-based targets to reduce GHG emissions in alignment with the Paris Agreement goals and global targets on ecosystems and biodiversity
   b. Establish full transparency and ban deforestation and other ecosystem conversion, crime, land grabs, and exploitation throughout supply chains
   c. Invest in land restoration strategies, such as the rehabilitation of degraded lands, planting trees, bushes and mangroves
   d. Develop true cost accounting for food
3. Influencing consumption
   a. Establish food loss and waste targets across the value chain
4. Ensuring inclusive development
   a. Implement commodity procurement strategy that favors long-term sustainable supply from equitable partnerships rather than buying on the spot market

This list is not exhaustive, nor are its components independent – for instance, developing true cost accounting for food can lead to innovative business models that more truly reflect the costs of farming on people and nature, thereby offering not only inclusive development but also transforming value chains. Given the scope of this sector and its reach within human systems, it is at once complex to tackle and capable of yielding high returns.

Opportunity related to economic goals

The Food and Land Use Coalition (FOLU) has estimated that there is an annual business opportunity of US$4.5 trillion associated with ten critical transitions to sustainably and climate-compatible agriculture in 2030 (FOLU, 2019, p.9). Their analysis breaks down this opportunity in the following way:

- Organic food and beverage ($770bn)
- Fortified food ($600bn)
- Product reformulation ($365bn)
- Dietary switch ($240bn)
- Tech in large scale and smallholder farms ($75bn)
- Forest restoration ($175bn)
- Sustainable aquaculture ($255bn)
- Plant-based meat ($140bn)
- Reducing food waste in the value chain ($120bn)
- Internet of things for agriculture ($110bn)
- Low-income food market ($270bn)
- Connectivity income gains ($130bn)

These calculations include savings that may accrue beyond companies mostly from a reduction in land use, reductions in food loss and waste. New opportunities will also arise from freed-up capital for investing into new assets in the new food and land use economy. In the final analysis, FOLU estimates that the total global contribution of transitioning this new economy, including reducing or eliminating current costs in food and land use systems, is estimated at $5.7 trillion a year by 2030 and $10.5 trillion by 2050.

These figures are staggering since they reflect the scale of the transformation that is necessary to achieve climate goals – and of the opportunity associated with it. Just as they are significant, there are barriers and risks associated with investing in these transformative measures. In general, these will be seen as risky, especially from the lens of finance. First, financing smallholders who tend to have limited collateral or high debt burdens will be seen as a risk; new business models, innovative technology, or new and unfamiliar practices that have not yet proved their financial viability or overall impact will also be perceived as risky. The geographies where these opportunities lie also tend to be in regions with high-risk profiles due to political risk, weaker legal systems, weak infrastructure (both institutional and physical) – all barriers to investment and at the very least factors that will slow down transformation. Hence while the sort of transformation related to agriculture and land use is consistent with companies' economic goals and sometimes can even represent a very profitable reorientation of their business, the barriers to transformation cannot be overlooked. It is very likely that most of these opportunities will stay on the table if those barriers remain unaddressed.

Opportunity related to strategic goals

Consequently, those companies that will work to weaken or overcome these barriers will benefit from a competitive advantage. Developing new business models and asset types for the new economy can have lower capital requirements, use fewer inputs and be overall more efficient. Businesses that can capture – and demonstrate – these advantages will not only unlock financing but lead in the transformation of the sector. Examples of such changes include alternative proteins that rely on new methods of farming, which require much less infrastructure and capital than livestock production. Regenerative methods of agriculture, too, require less resource input, specifically inorganic fertilizers, and pesticides. These leaner operation models represent a competitive advantage in a landscape of transformation so vast as that of agriculture and land use.

Furthermore, the innovation that will be unlocked in this shift will also distinguish leading companies. Solutions range from satellite-based technologies to support resource management, efficient irrigation management, digital farming, smart energy programs, and other innovations that have not yet been developed all have the potential to contribute to preserving natural resources, reducing GHG emissions, and attracting vital investment. Some of these innovations will also strengthen the resilience of smallholder farmers, by implication strengthening the resilience of value chains to which they belong.

Given the complexity of the agricultural and land use value chain, there is potential for small measures to have a ripple effect on entire sectors of value chains. Just like investment barriers, however, barriers to implementation are also high because of this very complexity. Similarly, it is through leadership and accepting higher risk levels that companies will reap the strategic benefits of agricultural transformation.

Opportunity related to environmental goals

Recalling that environmental goals are those that allow a company to secure access to the resources that are necessary to operate, produce, and deliver goods and services, the stakes are very high for this sector. Indeed, the impacts of climate change themselves affect both the quality and quantity of raw material, the infrastructure that allows for global distribution, and the conditions necessary for business continuity across the entire value chain.

The rewards of transforming the sector can be understood to be existential for the sector. Indeed, continuing to serve its purpose of feeding the global population and providing other natural services related to health and culture, for instance, will require companies at once to adapt to current and projected impacts and to find new ways to feed a growing world demos.

In terms of adaptation, the sector is already at the forefront of many innovative measures. Land restoration strategies, for instance, are key in this effort: the rehabilitation of degraded lands, but also planting trees, bushes, and mangroves, that can rebuild carbon sinks, increase crop yields, and improve soil and water quality while protecting biodiversity and preserving the ecosystem service could raise food production. Agroecology and agroforestry approaches also enhance carbon sequestration through the simultaneous combination of crop rotation and permanent soil cover. For instance, alternate wetting and drying of rice fields reduces methane emissions from paddies by 45% while saving water and producing yields similar to those of fully flooded rice (FAO, 2018, p.8). Furthermore, more and more measures to increase resilience are being documented and implemented, such as increasing plant biodiversity to break disease cycles (FAO, 2008), favoring natural fertility by increasing organic soil matter (IAEA, 2016), and so on.

Opportunity related to social goals

The opportunities of agriculture and land use range widely in terms of technical complexity, cost, or scale. However, given the deep-rootedness of the sector in many aspects of human life –nutrition, health, culture – and that it employs 28% of the global population (The World Bank, 2021), transforming the sector has far-reaching impacts on the lives and livelihoods of the vast majority of people in the world. In the developing world especially, where over 3 billion people, 80% of the poor and 40% of the world population, live in rural areas, with around 2.5 billion who depend on agriculture for their livelihoods (IFAD, 2017), this impact will be related to development goals.

The web of measures that will transform the sector will not only have an impact on emissions and adaptive capacity but will also build the resilience of these people and the communities in which they live. Greater access to information, technologies, and markets, for instance, which are all necessary to transform the sector, will not only contribute to designing production practices that are better from the point of view of climate. The other transformations that will result from these cannot all be charted but are sure to include greater resilience to other disruption, greater political and economic economy, more balanced leverage with partners in the supply chain, and generally improved welfare.

In addition, the transformation of the sector relies on the central role of farmers, pastoralists, fisherfolk, community forester, and teachers and specifically on the combination between their traditional knowledge and news skills. It is likely that, if rolled out successfully, this will have result in cultural shifts that value the collective knowledge of natural systems. This increased and improved repository of knowledge might further improve our collective capacity to respond to the impacts on climate change.

In 2019, the EAT-Lancet report linked sustainable agriculture, which will be central to the future of the planet and of its growing population, to healthier diets overall (Eat Forum, 2019). While averting emissions from livestock, protecting wildlife and biodiversity, and reducing the pollution of oceans, seas, and rivers, this kind of diet would also avoid an estimated 11 million deaths a year. By implication, without such changes, the burden of unhealth due to diet on the world would continue to grow and possibly worsen. There is a narrow self-interest to companies operating in the food sector to preserve the health of communities, who are also consumers and employees. There is, in addition, a broader interest companies may recognize that consists of not producing harmful goods or of trying to reduce the harm of those goods on those who consume them.

Finally, the role of women in food and agriculture across the planet cannot be overstated. Globally, they represent about 43% of the agricultural workforce, with this figure rising sharply to more than half in developing countries (Doss et al., 2011, p.1). What is more, they are powerful actors in shaping food systems due to their pivotal roles in decision making around nutrition, health, and family planning. By further ensuring that women are able to make these decisions with the best information and resources, and within contexts free of coercion – thanks to, for instance, equal access to land, labor, water, credit, political participation, financial services – their role in the transformation can become not only more central but more informed. Consequently, this transformation will likely be made more rapidly and in better conditions of success.

Conclusion

Market-wide initiatives such as the ones outlined here are already demonstrating a ripple effect on other sectors and global value chains, too, with other impacts beyond reducing emissions and building resilience. In particular, companies can integrate climate action without relinquishing other goals related to profits and growth, strategy, securing a supply chain, or securing the conditions for an able workforce. The economic opportunity associated with climate action has been documented by many observers and by companies themselves and, if availed of, could lead the necessary shift toward a decarbonized economy in line with science. That it is not can be attributed to a variety of barriers, both internal and external to companies.

Moreover, beyond such barriers lies the need for a new class of frameworks and economic models to capture the transformation that is needed to tackle the climate crisis, and that includes unprecedented technological innovation, a shift in how companies and societies more broadly relate to natural capital, the full health benefits of cleaner air, and other intangible consequences of the new economy. This book will aim to outline what is required to make this shift, but not before laying out the limits of current models in the next chapter.

References

Anuta, H., Ralon, P. and Taylor, M. (2018) *Renewable Power Generation Costs in 2018.* Available at: https://www.irena.org/-/media/Files/IRENA/Agency/Publication/2019/May/IRENA_Renewable-Power-Generations-Costs-in-2018.pdf (Accessed: 20 March 2021)

Ballentine, C. and Gupta, R. (2020) 'BlackRock Climate Vow Pays Off with ETF's $600 Million Debut,' *Bloomberg,* 11 February [Online]. Available at: https://www.bloomberg.com/news/articles/2020-02-11/blackrock-climate-vow-pays-off-with-new-etf-s-600-million-debut (Accessed: 20 March 2021)

Barry, C. (2018) 'Fiat Chrysler Unveils Plans to Make More Electrified Cars,' *AP News,* June 1 [Online]. Available at: https://apnews.com/article/ed200dcfd5684625954e90969ad3185c (Accessed: 20 March 2021)

BMW Group (2017) *BMW Group Announces Next Step in Electrification Strategy* [Press release]. 25 July. Available at: https://www.bmwgroup.com/content/dam/grpw/websites/bmwgroup_com/responsibility/downloads/en/2018/2017-BMW-Group-electrification-strategy-englisch.pdf (Accessed: 20 March 2021)

BNEF (2019) *Industrial Heat: Deep Decarbonization Opportunities.* Available at: https://about.bnef.com/ (Accessed: 20 March 2021)

Borlaug, B. et al. (2020) 'Levelized cost of charging electric vehicles in the United States,' *Joule,* 4(7), 1470–1485. [Online]. Available at: https://www.cell.com/joule/pdfExtended/S2542-4351(20)30231-2 (Accessed: 20 March 2021)

Brown, M. (2019) 'Solar Energy Prices Hit Tipping Point as China Reaches "Grid Parity",' *Inverse,* 11 February [Online]. Available at: https://www.inverse.com/article/58495-solar-energy-prices-hit-tipping-point-as-china-reaches-grid-parity (Accessed: 20 March 2021)

Buckland, K. and Tajitsu, N. (2019) 'Toyota Speeds up Electric Vehicle Schedule as Demand Heats up,' *Reuters,* June 7 [Online]. Available at: https://www.reuters.com/article/us-toyota-electric/toyota-speeds-up-electric-vehicle-schedule-as-demand-heats-up-idUSKCN1T806X (Accessed: 20 March 2021)

Carey, N. and White, J. (2018) 'Ford Plans $11 Billion Investment, 40 Electrified Vehicles by 2022,' *Reuters,* January 14 [Online]. Available at: https://www.reuters.com/article/us-autoshow-detroit-ford-motor/ford-plans-11-billion-investment-40-electrified-vehicles-by-2022-idUSKBN1F30YZ (Accessed: 20 March 2021)

Casale, M. and Mahoney, B. (2019). *Paying for Electric Buses Financing Tools for Cities and Agencies to Ditch Diesel.* Available at: https://uspirg.org/sites/pirg/files/reports/National%20-%20Paying%20for%20Electric%20Buses.pdf (Accessed: 20 March 2021)

Chassein, E., Roser, A. and John, F. (2017) *Using Renewable Energy for Heating and Cooling: Barriers and Drivers at Local Level.* Available at: http://www.progressheat.eu/IMG/pdf/progressheat_wp3.2_report_publication.pdf (Accessed: 20 March 2021)

Climate Bonds Initiative (2021) Climate Bonds Initiative [Online]. Available at: https://www.climatebonds.net/ (Accessed: 20 March 2021)

Daimler (2017) *Plans for More Than Ten Different All-Electric Vehicles by 2022: All Systems Are Go* [Online]. Daimler. Available at: https://media.daimler.com/marsMediaSite/en/instance/ko/Plans-for-more-than-ten-different-all-electric-vehicles-by-2022-All-systems-are-go.xhtml?oid=29779739 (Accessed: 20 March 2021)

DeBord, M. (2019) 'Ford Just Revealed Its Mustang Mach-E SUV – the First New Mustang-Branded Vehicle since 1964 and a Tesla Challenger,' *Business Insider,* November 18 [Online] Available at: https://www.businessinsider.com/ford-mustang-mach-e-revealed-los-angeles-photos-features-2019-11?IR=T (Accessed: 20 March 2021)

Decarb Heat (2020) *Decarb Heat Acting to Decarbonise Europe* [Online]. Decarb Heat. Available at: http://decarbheat.eu/ (Accessed: 25 May 2019)

DeLaski, A. and Mauer, J. (2017) *Energy-Saving States of America: How Every State Benefits from National Appliance Standards*. ASAP and ACEEE Working Paper. Available at: https://appliance-standards.org/sites/default/files/Appliances%20standards%20white%20paper%202%202-14-17.pdf (Accessed: 20 March 2021)

Doss, C. et al. (2011) *The Role of Women in Agriculture*. ESA Working Paper No. 11-02. [Online]. Available at: http://www.fao.org/3/am307e/am307e00.pdf (Accessed: 20 March 2021)

E2, E4 The Future (2019) *Energy Efficiency Jobs in America*. Available at: https://www.e2.org/wp-content/uploads/2019/09/Energy-Efficiency-Jobs-in-America-2019-Full-Report.pdf (Accessed: 20 March 2021)

Eat Forum (2019) *The EAT-Lancet Commission on Food, Planet, Health* [Online]. Available at: https://eatforum.org/eat-lancet-commission/ (Accessed: 20 March 2021)

Edmond, C. (2019) 'China's Lead in the Global Solar Race - at a Glance,' *World Economic Forum*, 19 June [Online]. Available at: https://www.weforum.org/agenda/2019/06/chinas-lead-in-the-global-solar-race-at-a-glance/ (Accessed: 20 March 2021)

EEA, European Environment Agency (2019) *Emissions of Air Pollutants from Transport* [Online]. EEA. Available at: https://www.eea.europa.eu/data-and-maps/indicators/transport-emissions-of-air-pollutants-8/transport-emissions-of-air-pollutants-8 (Accessed: 20 March 2021)

EIA (2021a) Table 11.5 Carbon Dioxide Emissions from Energy Consumption: Transportation Sector [Online]. EIA. Available at: https://www.eia.gov/totalenergy/data/monthly/pdf/sec11_8.pdf (Accessed: 20 March 2021)

EIA (2021b) Wind Explained Electricity Generation from Wind. [Online]. EIA. Available at: https://www.eia.gov/energyexplained/wind/electricity-generation-from-wind.php#:~:text=U.S.%20total%20annual%20Electricity%20generation,U.S.%20utility%2Dscale%20electricity%20generation (Accessed: 20 March 2021)

EESI, Environmental and Energy Study Institute (2019) 'Fact Sheet | Jobs in Renewable Energy, Energy Efficiency, and Resilience (2019),' *EESI*, 23 July [Online] Available at: https://www.eesi.org/papers/view/fact-sheet-jobs-in-renewable-energy-energy-efficiency-and-resilience-2019 (Accessed: 20 March 2021)

Energy2030 (2015) *Energy2030* [Online]. Available at: http://www.energy2030.org/ (Accessed: 20 March 2021)

EPA (2020) Sources of Greenhouse Gas Emissions [Online]. EPA. Available at: https://www.epa.gov/ghgemissions/sources-greenhouse-gas-emissions (Accessed: 20 March 2021)

EY (2019) *Renewable Energy Is Taking Strides towards a Subsidy-Free Era, EY Report Reveals* [Press release]. 15 May. Available at: https://www.ey.com/en_gl/news/2019/05/renewable-energy-is-taking-strides-towards-a-subsidy-free-era-ey-report-reveals (Accessed: 20 March 2021)

FAO (2008) NSP - Biodiversity and Ecosystem Services. [Online]. FAO. Available at: http://www.fao.org/agriculture/crops/thematic-sitemap/theme/biodiversity0/en/ (Accessed: 20 March 2021)

FAO (2018) *Rice Landscapes and Climate Change*. Available at: http://www.fao.org/3/CA3269EN/ca3269en.pdf (Accessed: 20 March 2021)

FOLU - The Food and Land Use Coalition (2019) *Growing Better: Ten Critical Transitions to Transform Food and Land Use*. Available at: https://www.foodandlandusecoalition.org/wp-content/uploads/2019/09/FOLU-GrowingBetter-GlobalReport.pdf (Accessed: 20 March 2021)

Garrett-Peltier, H. (2017) 'Green versus brown: Comparing the employment impacts of energy efficiency, renewable energy, and fossil fuels using an input-output model,' *Economic Modelling*, 61, 439–467. [Online]. Available at: https://doi.org/10.1016/j.econmod.2016.11.012 (Accessed: 20 March 2021)

Gimon, E. et al. (2019) *The Coal Cost Crossover: Economic Viability of Existing Coal Compared to New Local Wind and Solar Resources*. Available at: https://energyinnovation.org/wp-content/uploads/2019/04/Coal-Cost-Crossover_Energy-Innovation_VCE_FINAL2.pdf (Accessed: 20 March 2021)

Hall, D. and Lutsey, N. (2019) 'Estimating the Infrastructure Needs and Costs for the Launch of Zero-Emission Trucks,' *ICCT*, 9 August [Online] Available at: https://theicct.org/publications/zero-emission-truck-infrastructure (Accessed: 20 March 2021)

Hawkins, A.J. (2019) 'Volvo Unveils Its First Fully Electric Car – and a Bold Pledge to Go Carbon-Neutral,' *The Verge*, 16 October [Online]. Available at: https://www.theverge.com/2019/10/16/20915841/volvo-xc40-recharge-electric-suv-specs-miles-range-reveal (Accessed: 20 March 2021)

Heat Roadmap (2020) *Heat Roadmap Europe* [Online]. Available at: https://heatroadmap.eu/ (Accessed: 25 May 2019)

Honda (2019) *Honda Commits to Total Electrification in Europe by 2025* [Press release]. 05 March. Available at: https://hondanews.eu/eu/lt/corporate/media/pressreleases/162386/honda-commits-to-total-electrification-in-europe-by-2025 (Accessed: 20 March 2021)

IAEA (2016) Improving Soil Fertility [Online]. IAEA. Available at: https://www.iaea.org/topics/improving-soil-fertility (Accessed: 20 March 2021)

IEA (2018) *Renewable Heat Policies*. Available at: https://www.iea.org/reports/renewable-heat-policies (Accessed: 20 March 2021)

IEA (2020) Statistics Report Energy Efficiency Indicators Highlights. Available at: https://www.iea.org/reports/energy-efficiency-indicators (Accessed: 20 March 2021)

IFAD (2017) Why Rural People? [Online]. IFAD. Available at: https://www.ifad.org/en/investing-in-rural-people (Accessed: 20 March 2021)

IPCC (2014) *Climate Change 2014 Mitigation of Climate Change*. Available at: https://archive.ipcc.ch/pdf/assessment-report/ar5/wg3/ipcc_wg3_ar5_chapter1.pdf (Accessed: 20 March 2021)

IPCC (2020) *Climate Change and Land*. Available at: https://www.ipcc.ch/site/assets/uploads/sites/4/2020/02/SPM_Updated-Jan20.pdf (Accessed: 20 March 2021)

IRENA (2018) Trends in Renewable Energy [Online]. Public Tableau. Available at: https://public.tableau.com/views/IRENARETimeSeries/Charts?%3Aembed=y&%3AshowVizHome=no&publish=yes&%3Atoolbar=no (Accessed: 20 March 2021)

IRENA (2019) *Renewable Energy and Jobs – Annual Review 2019*. Available at: https://www.irena.org/publications/2019/Jun/Renewable-Energy-and-Jobs-Annual-Review-2019 (Accessed: 20 March 2021)

IRENA and European Commission (2018) *Renewable Energy Prospects for the European Union*. Available at: https://www.irena.org/-/media/Files/IRENA/Agency/Publication/2018/Feb/IRENA_REmap_EU_2018.pdf (Accessed: 20 March 2021)

IRENA, IEA and REN21 (2018) *Renewable Energy Policies in a Time of Transition*. Available at: https://www.irena.org/publications/2018/Apr/Renewable-energy-policies-in-a-time-of-transition (Accessed: 20 March 2021)

Jacobs, M. and Mountford, H. (2015) *The 2015 New Climate Economy Report. Seizing the Global Opportunity. Partnerships for Better Growth and A Better Climate*. Available at: http://newclimateeconomy.report/2015/wp-content/uploads/sites/3/2014/08/NCE-2015_Seizing-the-Global-Opportunity_web.pdf (Accessed: 20 March 2021)

Jones, L. (2021) 'Record $269.5bn Green Issuance for 2020: Late Surge Sees Pandemic Year Pip 2019 Total by $3bn,' *Climate Bonds Initiative*, 24 January [Online]. Available at: https://www.climatebonds.net/2021/01/record-2695bn-green-issuance-2020-late-surge-sees-pandemic-year-pip-2019-total-3bn (Accessed: 20 March 2021)

Joselow, M. (2018) 'The U.S. Has 1 Million Electric Vehicles, but Does It Matter?,' *Scientific American*, 12 October [Online]. Available at: https://www.scientificamerican.com/article/the-u-s-has-1-million-electric-vehicles-but-does-it-matter/ (Accessed: 20 March 2021)

Knudsen, C. and Doyle, A. (2019) 'Norway's Electric Cars Zip to New Record: Almost a Third of All Sales,' *Reuters*, January 2 [Online]. Available at: https://www.reuters.com/article/us-norway-autos/norways-electric-cars-zip-to-new-record-almost-a-third-of-all-sales-idUSKCN1OW0YP (Accessed: 20 March 2021)

Krupnick A.J., Rowe R.D. and Lang C.M. (1997) 'Transportation and Air Pollution: The Environmental Damages' in Greene D.L., Jones D.W. and Delucchi M.A. (eds.) *The Full Costs and Benefits of Transportation*. New York: Springer [Online]. Available at: https://link.springer.com/chapter/10.1007/978-3-642-59064-1_12#citeas (Accessed: 20 March 2021)

Lazard (2019) 'Levelized Cost of Energy and Levelized Cost of Storage 2019,' *Lazard*, 7 November [Online]. Available at: https://www.lazard.com/perspective/lcoe2019 (Accessed: 20 March 2021)

Le Quéré, C. et al. (2020) 'Temporary reduction in daily global CO_2 emissions during the COVID-19 forced confinement', *Nature Climate Change*, 10 [Online]. Available at: https://www.nature.com/articles/s41558-020-0797-x (Accessed: 20 March 2021)

LeasePlan (2019) *The Total Cost Of Ownership of Electric Vehicles Compared to Traditional Vehicles*. Available at: https://www.leaseplan.com/en-ix/blog/tco/tco-ev/ (Accessed: 20 March 2021)

Lutsey, N. and Nicholas, M. (2019) *Update on Electric Vehicle Costs in the United States through 2030* [Online]. Available at: https://theicct.org/sites/default/files/publications/EV_cost_2020_2030_20190401.pdf (Accessed: 20 March 2021)

Markets Insider (2019) *Sourcing Blockchain Network' to Help Build a More Ethical Electric Vehicle Supply Chain* [Press release]. 10 December. Available at: https://markets.businessinsider.com/news/stocks/fca-joins-responsible-sourcing-blockchain-network-to-help-build-a-more-ethical-electric-vehicle-supply-chain-1028751836 (Accessed: 20 March 2021)

Massa, A. (2020) 'BlackRock Puts Climate at Center of $7 Trillion Strategy,' *Bloomberg*, 14 January [Online]. Available at: https://www.bloomberg.com/news/articles/2020-01-14/blackrock-puts-environmental-sustainability-center-of-strategy (Accessed: 20 March 2021)

Matousek, M. (2019) 'Electric Vehicles Are a Tiny Piece of the Global Car Market, but Volkswagen Is Making a Huge Bet on Them. It Doesn't Have a Choice,' *Business Insider*, 8 November [Online]. Available at: https://www.businessinsider.fr/us/vw-making-huge-bet-on-electric-vehicles-in-next-decade-2019-11 (Accessed: 20 March 2021)

McKinsey & Company (2020) *Electric Mobility after the Crisis: Why An Auto Slowdown Won't Hurt EV Demand*. Available at: https://www.mckinsey.com/~/media/McKinsey/Industries/Automotive%20and%20Assembly/Our%20Insights/Electric%20mobility%20after%20the%20crisis%20Why%20an%20auto%20slowdown%20wont%20hurt%20EV%20demand/Electric-mobility-after-the-crisis-Why-an-auto-slowdown-wont-hurt-EV-demand.pdf?shouldIndex=false (Accessed: 20 March 2021)

Mountford, H. et al. (2018) *Unlocking the Inclusive Growth Story of the 21st Century: Accelerating Climate Action in Urgent*. Available at: https://newclimateeconomy.report/2018/wp-content/uploads/sites/6/2019/04/NCE_2018Report_Full_FINAL.pdf (Accessed: 20 March 2021)

Muro, M. et al. (2019) *Advancing Inclusion through Clean Energy Jobs*. Available at: https://www.brookings.edu/research/advancing-inclusion-through-clean-energy-jobs/ (Accessed: 20 March 2021)

RE100 (2018) *RE100 Progress and Insights Annual Report, November 2018*. Moving to truly global impact. Influencing renewable electricity markets. Available at: http://media.virbcdn.com/files/fd/868ace70d5d2f590-RE100ProgressandInsightsAnnualReportNovember2018.pdf (Accessed: 20 March 2021)

RMI (2019) *Report The Growing Market for Clean Energy Portfolios*. Available at: https://rmi.org/insight/clean-energy-portfolios-pipelines-and-plants/ (Accessed: 20 March 2021)

Szymkowski, S. (2019) 'Every Honda Sold in Europe Will Be Electrified by 2022,' *Road Show by CNET*, 23 October [Online]. Available at: https://www.cnet.com/roadshow/news/honda-electrified-europe/ (Accessed: 20 March 2021)

Taylor, M. and Young So, E. (2016) *Solar PV in Africa: Costs and Markets*. Available at: https://www.irena.org/-/media/Files/IRENA/Agency/Publication/2016/IRENA_Solar_PV_Costs_Africa_2016.pdf (Accessed: 20 March 2021)

Teter, J. (2020) *Tracking Transport 2020*. Available at: https://www.iea.org/reports/tracking-transport-2020 (Accessed: 20 March 2021)

The Financial Stability Board Task Force on Climate-Related Financial Disclosures (2020) *Task Force on Climate-Related Financial Disclosures 2020 Status Report*. Available at: https://www.fsb.org/wp-content/uploads/P291020-1.pdf (Accessed: 20 March 2021)

The New Climate Economy (2016) *Infrastructure Investment Needs of a Low-Carbon Scenario*. Available at: https://newclimateeconomy.report/workingpapers/wp-content/uploads/sites/5/2016/04/Infrastructure-investment-needs-of-a-low-carbon-scenario.pdf (Accessed: 20 March 2021)

The World Bank (2021) Employment in Agriculture (% of Total Employment) (Modeled ILO Estimate) [Online]. The World Bank. Available at: https://data.worldbank.org/indicator/SL.AGR.EMPL.ZS (Accessed: 20 March 2021)

Turk, D. and Kamiya, G. (2021) 'The Impact of the Covid-19 Crisis on Clean Energy Progress,' *IEA*, 11 June [Online]. Available at: https://www.iea.org/articles/the-impact-of-the-covid-19-crisis-on-clean-energy-progress (Accessed: 20 March 2021)

UNFCCC (2018) *100+ cities Produce More than 70% of Electricity from Renewables - CDP* [Press release]. 27 February. Available at: https://unfccc.int/news/100-cities-produce-more-than-70-of-electricity-from-renewables-cdp (Accessed: 20 March 2021)

Volkswagen (2019) *Volkswagen Significantly Raises Electric Car Production Forecast for 2025* [Press release]. 27 December. Available at: https://www.volkswagen-newsroom.com/en/press-releases/volkswagen-significantly-raises-electric-car-production-forecast-for-2025-5696 (Accessed: 20 March 2021)

Volvo (2019a) *The Future is Electric* [Online]. Volvo Cars. Available at: https://group.volvocars.com/company/innovation/electrification (Accessed: 20 March 2021)

Volvo (2019b) *Volvo Cars to Radically Reduce Carbon Emissions as Part of New Ambitious Climate Plan* [Press release]. 16 October. Available at: https://www.media.volvocars.com/global/en-gb/media/pressreleases/259147/volvo-cars-to-radically-reduce-carbon-emissions-as-part-of-new-ambitious-climate-plan (Accessed: 20 March 2021)

White, J. (2019) 'GM Says Most Cadillac Vehicles Will Be Electric within the Next 10 Years,' *Business Insider*, December 13 [Online]. Available at: https://www.businessinsider.com/gm-says-cadillac-vehicles-will-be-all-electric-by-2030-2019-12?IR=T (Accessed: 20 March 2021)

WHO (2019) Air Pollution [Online]. WHO. Available at: https://www.who.int/health-topics/air-pollution#tab=tab_1 (Accessed: 20 March 2021)

Yan, J. (2019) 'City-level analysis of subsidy-free solar photovoltaic electricity price, profits and grid parity in China,' *Nature Energy*, 4 [Online]. Available at: https://www.nature.com/articles/s41560-019-0441-z (Accessed: 20 March 2021)

Zhang, B. (2018) 'We Drove a New $38,000 Nissan Leaf to See How It Stacks Up against Tesla and the Chevy Bolt – Here's the Verdict,' *Business Insider*, May 9 [Online]. Available at: https://www.businessinsider.com/nissan-leaf-ev-review-new-generation-2018-3?IR=T (Accessed: 20 March 2021)

PART II
The current corporate climate leadership

4
BUILDING THE CLIMATE-COMPATIBLE ECONOMY

Strategies for corporate decarbonization and resilience

In January 2021, a group of the world's leading natural scientists published a short and stark overview of the interrelated global risks facing humanity over the coming decades. Writing candidly, they suggested "the choice before us is between exiting overshoot by design or disaster – because exiting overshoot is inevitable one way or another".[1]

Exiting overshoot by design will require a new corporate climate leadership with three components. First, companies must decarbonize to avoid unmanageable climate change. This means reducing greenhouse gas emissions by 45% below 2010 levels by 2030 and achieving net-zero emissions by 2050. Second, companies must build socio-ecological resilience to manage the unavoidable climate impacts we are already experiencing. This means investing in the six so-called capital assets – human, social, natural, physical, financial, and political – that are the key building blocks of resilience. Third, companies must create a shared prosperity by building a just transition from the high-carbon and deeply unequal society of today to the new economy of tomorrow. This chapter presents the innovations that companies can use to decarbonize energy, transport, and food systems; guidance on how to apply the six capital assets inside individual companies, across complex global supply chains, and within frontline communities; and advice on how to secure a just transition.

Avoid the unmanageable: decarbonizing the real economy

There is an old adage in climate science that greenhouse gas pollution is predominantly "food, fuel and footnotes". This is certainly backed up by data. In 2016, total global greenhouse gas emissions reached 49.4 billion tons of CO_2 equivalents (CO_2e), of which 73.2% of GHGs came from fuel (57% from energy use and 16.2% from transportation); while 18.2% came from agriculture and

DOI: 10.4324/9781003025948-7

land use.[2] Decarbonizing the economy requires reducing GHG emissions by 45% by 2030,[3] net-zero carbon dioxide (CO_2) emissions by 2050, and total GHG emissions reaching net-zero between 2063 and 2068.[4] The concept of net-zero involves addressing both "sources" and "sinks". Human-caused emissions should be reduced to as close to zero as possible. Any remaining GHGs should be addressed through carbon removal, for example, by restoring forests. Net-zero emissions are achieved when any remaining human-caused GHG emissions are balanced out by removing GHGs from the atmosphere. This section explores how companies can achieve these sizable emissions reductions targets predominantly through innovations in energy, transport, and agriculture.

Significant emissions reductions can be achieved in energy systems through investments in solar, wind turbines, geothermal, microgrids, wave and tidal, biomass, nuclear, cogeneration, in-stream hydro, waste-to-energy, energy storage, energy efficiency, and energy efficiency.

Reductions from land-use and agriculture value chains can be achieved by changing food consumption from meat to plant-rich diets; reducing food waste; farmland restoration; agroforestry, nutrient management; composting; forest protection including afforestation; and restoration of wetlands.

Decarbonization in the transport sector requires modal shifts including improving public transport; replacing the internal combustion engine with electric vehicles; building walkable cities with bike infrastructure; expanding high-speed rail; improving logistics; and the deployment of innovations to both maritime and aviation transportation.

Other notable emissions reductions can be achieved by investing in net-zero buildings, recycling, innovations in cement, and changes in technologies dealing with refrigeration and bio-materials. This section explores these readily available innovations in more detail.

Food: reducing emissions across the food and agriculture value chains

The Intergovernmental Panel on Climate Change (IPCC) defines the global food system as all the elements (environment, people, inputs, processes, infrastructures, institutions, etc.) and activities that relate to the production, processing, distribution, preparation, and consumption of food. The food system involves supply (production, processing, marketing, and retailing) and demand (consumption and diets). Food system drivers (ecosystem services, economics and technology, social and cultural norms and traditions, and demographics) combine with the enabling conditions (policies, institutions, and governance) to affect food security, nutrition and health, livelihoods, economic benefits, and environmental effects including climate change.[5]

The global food system is responsible for 37% of all greenhouse gas emissions. The amount of emissions, and the share of the overall total, could increase over the coming decades depending on the resolution of a number of trends such as[6] consumption trends arising from a growing global population; the growth of the global middle class; changing dietary habits including a preference for meat; the conversion of natural ecosystems into managed land; rapid urbanization; and intensification of land management. These could all drive emissions up. In addition, poorly crafted emissions reductions responses, such as biofuels which require large land areas, could compete with existing uses of land and drive emissions through expanded uses of fertilizer.[7] Climate change may also undermine the role of land as a sink, reducing its current capacity to absorb and sequester up to 29% of total CO_2 emissions and therefore undermining efforts to achieve net-zero and limit global warming to 1.5°C.[8]

Companies can work toward decarbonizing food systems by reducing emissions at source or through enhancing carbon sinks through Carbon Dioxide Removal (CDR). The IPCC has identified more than 60 response options across land management; value-chain management; risk management; and enabling responses (predominantly financial and policy instruments) that can guide companies toward the decarbonization goal. The largest potential for reducing emissions are reduced deforestation and forest degradation; a shift toward plant-based diets; and reduced food and agricultural waste. The options with largest CDR potential are afforestation/reforestation; soil carbon sequestration in croplands and grasslands; and bioenergy with carbon capture and storage.[9] These are explored below.

Sustainable land management (SLM) describes the stewardship of soils, water, animals and plants, to advance food security, health, and economic development while addressing climate change and other localized environmental degradation. SLM includes management of cropland and grazing lands; forest management; reduced deforestation and degradation; ecosystem conservation and land restoration; and increased soil organic carbon content. For example, agricultural soils have lost 20%–60% of their soil carbon from tillage and harvesting and the change from deep-rooted perennial plants to shallow-rooted annual plants. The mitigation potential for soil organic matter stocks is equivalent to almost 20% of global GHGs. Companies in the food value chain can enhance the capacity of soils to absorb carbon, delivering mitigation potential of almost 20% of global GHGs, by insisting on tried and tested practices including minimum tillage, crop residue retention and soil surface coverage, and crop rotations.[9]

Value-chain management includes sustainable food production; increased food productivity; dietary choices; and reduced food loss and waste. Approximately 25%–30% of total food produced is lost or wasted. Reducing food waste by 50% would generate net emissions as high as 30% of total food-sourced GHGs, equivalent to more than 10% of annual emissions and slightly more than the GHG emissions of the entire European Union.[9] Companies – both those operating within the food chain and others with a large "food footprint" (large cafeterias, for instance) have a significant role to play in reducing food waste. They can conduct food waste audits to determine how and why they waste food and identify opportunities to improve their performance. They can foster better communication across the food value chain to match demand and supply as discrepancies between demand and supply are a major cause of food wastage. This could include optimizing production processes and cold chain management to minimize losses during production and transport and to improve purchasing and inventory practices to reduce over-ordering and spoilage.[10] Retailers can move away from the practice of rejection of food products on the basis of aesthetic concerns. The FAO points out that farmers discard between 20% and 40% of their fresh produce because it doesn't meet retailer's cosmetic specifications.[11] Finally, food companies could reduce the amount of food sent to landfills by directing food scraps to agricultural partners and educate customers about proper food storage, preparation, and disposal methods.

Shifting to diets that are lower in emissions-intensive foods like beef delivers GHG reductions equivalent to the whole of North America. Such high emissions stem from the resource-intensive production of meat. For beef cattle, this conversion ratio is as high as 43:1, meaning that 43 kg of feeder is needed to produce 1 kg of beef product. Furthermore, emissions from the animal itself through manure and digestion contribute to the high emissions of animal-based products.[12] Decreasing meat consumption further reduces water use, soil degradation, pressure on forests, and land used for feed, potentially freeing up land for mitigation.[13]

Risk management and enabling responses include insurance; early warning systems; seed sovereignty; livelihood diversification; prohibitions against nitrate pollution or soil erosion; environmental quality standards; and policy instruments to intervene in markets such as carbon pricing or green payments. For example, Allianz Re was critical to the development of a multi-stakeholder partnership to provide governments and NGOs with better information on rice crop growth to support the development of more robust food security policies in addition to new and enhanced crop insurance programs in Southeast Asia. The Remote Sensing-based Information and Insurance for Crops in Emerging Economies (RIICE) initiative increases the availability and quality of information on rice yields to help improve management of domestic rice production and distribution, especially after extreme events. It also provides access to insurance solutions to alleviate the financial effects on farmers that stem from natural catastrophes such as flood and drought. Radar images are used to determine how much rice grows in each area, each season, ultimately arriving at a total national yield estimate. By analyzing time series, RIICE determines the extent of rice cropping, monitors rice growth, and identifies crop damages and losses caused by droughts and floods. The data captured by the satellites is processed and then translated into readable maps.[14]

Fuel: decarbonization in the energy and transport sectors

Decarbonization of the energy sector requires greater focus on energy conservation, efficiency, improved transmission, and switching from fossil fuels to renewable sources such as wind and solar. A team of researchers has produced roadmaps for 139 countries that transform energy infrastructures consistent with net-zero emissions, predominantly relying on wind and solar.[15]

Wind and solar are both essential to achieving net-zero GHG emissions by 2050. Wind turbines provide the lowest-cost source of new renewable energy capacity. The wind energy potential of Kansas, North Dakota, and Texas would be meet all electricity demand in the US. Wind farms use less than 1% of the land they sit on, so grazing, farming, recreation, or conservation can happen simultaneously with power generation. It takes one year or less to build a wind farm, quickly producing energy and return on investment.[16]

Solar farms operate at utility scale. Across their lifecycle they emit 94% less CO_2 emissions than coal plants. Utility-scale solar PV currently accounts for .4% of global electricity generation. Raising this to 10% could help avoid over 36 Gt of CO_2, equivalent to the amount of GHGs emitted by the whole emissions from fossil fuel energy sources in 2010,[17] while saving $5 trillion in operational costs by 2050. In addition, rooftop solar and other forms of onsite energy generation avoid losses in grid transmission and, when combined with net-metering that allows excess electricity to be sold back to the grid, can make solar panels financially feasible by introducing new revenue streams.[18]

Numerous other energy options contribute to decarbonization. For example, 39 countries could supply 100% of their electricity needs from geothermal energy. Some projections indicate that undiscovered geothermal could supply up to 13% of global current electricity needs.[18] Nuclear power has many legitimate reasons to be controversial, including issues around safety, security, waste, and cost. However, the IPCC estimates that net-zero emissions will be out of reach without some nuclear in the energy mix. This is because nuclear power has a low carbon footprint, with GHG emissions between 10 and 100 times lower than coal.[19] Increasing the use of microgrids is essential because they reduce energy lost in transmission and distribution, increasing energy efficiency compared to a centralized grid. According to the Lawrence Livermore Laboratory over 66% of all electricity generation in the United States is rejected, meaning lost through waste heat due to inefficiencies in generation, transmission and distribution right from the generating plant through to end use in businesses and homes.[20]

The transport sector is responsible for 23% of global GHG emissions, and two-thirds of the world's oil consumption is used for fuel in cars and trucks.[18] Road freight is accountable for about 6% of all emissions worldwide, with trucks transporting nearly 70% of all domestic freight tonnage in the United States and consuming more than 25% of the fuel. The causes rest with the inefficiencies in the internal combustion engine. Only 21% of the energy consumption in conventional cars and trucks propel them forward, as opposed to 60% in an electric vehicle (EV). If adoption of fuel-saving technologies grows from 2% to 85% of trucks by 2050, there would be a reduction of 6.2 gigatons of CO_2 and save $2.8 trillion on fuel costs over thirty years.[18] As a result, more companies should follow the model announced by Walmart[21] and Amazon[22] to move toward fully electric vehicle fleets.

Maritime transport produces 3% of global GHG emissions. By weight, more than 80% of global trade is transported by vessels. Improvements in design, technology, maintenance, and operations can increase shipping efficiency, which can have a significant impact on reducing emissions and improving air quality and human health. Available efficiency approaches can reduce shipping emissions by 30%–55% by 2030. If there is an efficiency gain of 50% across the international shipping industry by 2050, 7.9 gigatons of CO_2 emissions can be avoided. $424 billion could be saved in fuel costs over thirty years and $1 trillion over the life of the ships.[18] Companies can contribute to reducing emissions in the maritime sector by aligning with the GLEC Framework. Developed by the Smart Freight Centre (SFC), the Framework is the only global method for total supply chain carbon accounting, enabling companies to consistently calculate their GHG footprint across the global multi-modal supply chain and inform their business decisions and efforts to reduce emissions.[23]

Footnotes: additional sectors and measures to drive toward net-zero emissions by 2050

Carbon dioxide (CO_2) is the most common of the six greenhouse gas emissions. These gases have different warming potential and consequently different implications for the climate system. For example, Methane is 28 times more powerful than CO_2 despite being less prevalent in the atmosphere. Landfills are a top source of methane emissions, releasing 12% of the world's total. Landfill methane can be tapped and used as a fairly clean energy source instead of being left to leak into the air of left in the waste.[18] Worldwide, buildings account for 19% of GHG emissions. In the U.S., buildings' energy consumption is greater than 40% of the nation's total. Retrofitting existing residential and commercial building space could reduce US emissions by 10%, lead to over $1 trillion in energy savings over a ten-year period, and create the equivalent of 3.3 million cumulative job years of employment.[18]

Household and Industrial Recycling can reduce resource extraction by reusing materials rather than producing new products. Waste production multiplied tenfold over the past century, and it is predicted to double again by 2025. About 50% of recycled products come from households, and the other 50% comes from industrial and commercial sectors. If the average worldwide recycling rate increases to 65% of total recyclable waste, commercial and industrial sectors can avoid 2.8 of CO_2 emissions by 2050.[18]

In addition to reducing emissions at source, removing GHGs from the atmosphere using sinks and sequestration is a vital part of the net-zero journey. Carbon removal can be achieved by enhancing carbon sinks, using natural strategies like afforestation, agricultural soil management, and marine protected areas and expanding carbon sequestration such as direct air capture and storage (DACS) technology, carbon capture and storage, and enhanced mineralization. For example, expanding, restoring, and managing forests can leverage the power of photosynthesis, converting carbon dioxide in the air into carbon stored in wood and soils. The World Resources Institute (WRI) estimates that the carbon-removal potential from forests in the United States alone is equivalent to all annual emissions from the US agricultural sector, is inexpensive compared to other carbon removal options, and can deliver co-benefits such as cleaner water and air.[24]

Ultimately, decarbonization does not rest with one silver bullet technology. Instead, reaching net-zero emissions by 2050 requires a broad range of choices across every sector of the economy. The IPCC has identified more than 150 separate innovations in policy, technology, finance, and industrial processes that can reduce greenhouse gas emissions. The challenge and opportunity for each company are to find the right combination of those innovations to suit their own particular business culture and economic sector. Each innovation is available today and is cost effective.[17]

Manage the unavoidable: enhancing socio-ecological resilience

The IPCC defines resilience as

> the ability of a system and its component parts to anticipate, absorb, accommodate, or recover from the effects of a hazardous event in a timely and efficient manner, including through ensuring the preservation, restoration, or improvement of its essential basic structures and functions.[25]

A resilient business must be able to anticipate, absorb, accommodate and rapidly recover from climate events. Business continuity requires these abilities to be present within own operations, throughout the supply chain, and within frontline communities. A complex, global and interconnected business cannot be resilient if it focuses exclusively on efforts within its own four walls. It needs to reach out to moderate harm to socio-ecological systems and enable people, the economy, and natural systems to rebound quickly in the face of adversity. Businesses can be agents of climate resilience, benefitting from the availability of resources; the security of supply chains and transport routes; the protection of workers and infrastructure; and the rising prosperity of consumers and shareholders.[26]

Investing in six so-called Capital Assets represents the most effective and comprehensive means for the private sector to build resilience. These are interdependent capacities that, together, address the underlying causes of vulnerability such as poverty, inequality, and environmental degradation.[27] The capital assets can also reduce exposure to climate risk, notably by improving poor planning and construction practices that currently place infrastructure, population centers, and utilities in the path of climate hazards, and often without sufficient regard for how these hazards are increasing in intensity and frequency.

Human capital

Human capital refers to the skills and knowledge of available human resources, particularly in the workforce. A company might enhance human capital by investing in skills and training to cultivate agents of broader workplace, household, and community resilience. It might lead on technology development, transfer, and diffusion; conduct hazard, exposure, and vulnerability mapping focused on its workforce; work with government to produce early-warning and early action to an impending severe weather event for their communities; and undertake participatory scenario development to prepare workers for climate impacts. Companies can also enhance human capital by revisiting their employment practices. For example, by enhancing access to their jobs through their diversity and inclusion programs that seek to bring marginalized groups such as refugees, the homeless, or the formerly incarcerated into the workplace. They can further think about employment access when deciding where to situate their businesses, for example, by deciding to locate their facilities near public transportation or in regions of a country with high unemployment.[28]

Increasingly, investments in human capital involve advancing gender equality. Women possess significant knowledge and skills that can contribute to effective responses to climate change, including building community resilience, yet they are often not empowered to meaningfully participate in decision-making processes. Their unique knowledge of community dynamics and skills in the use and management of natural resources can enhance the efficiency and sustainability of climate change response efforts.

Social capital

Social capital refers to strong relationships, collaborations, and bonds of mutual support and cooperation that are essential for addressing a systematic global challenge such as climate change. When reciprocal claims for support can be made within communities in times of stress, this adds considerably to adaptive capacity. Activities and businesses that strengthen social bonds and aid the spread of ideas and resources are considered extremely important elements of social capital. A company might enhance social capital by establishing planning boards designed to evaluate risk and create strategies for resilience. These boards should include worker representatives. Social media and technology companies might work to enhance virtual social networks that can provide support in times of crisis.[28] Companies that help to build social networks are also contributing to social capital. Finally, collaborative initiatives that build cooperative relationships across supply chains, amongst peer companies and with other stakeholder groups, also contribute to social capital.

Social protection programs are also vital elements of social capital. These include policies intending to reduce poverty, deprivation, and vulnerability and may include social safety nets (cash and food transfers, public works programs, school feeding programs); social insurance (pensions, health or unemployment benefits); labor market interventions (job benefits, labor standards); and social care services (for the elderly or disabled). The principal lesson from the COVID-19 crisis should be that vulnerability is exacerbated by the absence of adequate social protection. Zero-hour contracts and lack of sick-pay force people who are ill to work and prevent them from building a financial reserve upon which to draw in times of crisis. Lack of health coverage prevents people from seeking medical treatment when they are ill, with the potential for spreading rather than containing disease. Insufficient funding of unemployment and pension benefits forces people into destitution in times of economic hardship. Companies should therefore look at how they can expand social protection inside the company through their own employment and benefits practices while advocating for public policy to support strong social safety nets.

Natural capital

Natural capital refers to the full range of services provided by biodiversity and ecosystem services, including land and water. For example, wetlands are vital to climate resilience because they protect upland areas – including valuable residential and commercial property – from flooding due to sea-level rise and storms; and help to regulate water tables. They further prevent coastline erosion due to their ability to absorb the energy created by ocean currents. According to research published by the Royal Swedish Academy of Sciences, coastal wetlands reduce the damaging effects of hurricanes on coastal communities with significant financial benefits. A regression model using 34 major US hurricanes since 1980 determined that coastal wetlands in the US currently provide over $23 billion per year in storm protection services as these wetlands function as self-maintaining "horizontal levees".[29]

Companies might work to enhance natural capital by maintaining wetlands and urban green spaces; expanding forested areas, which help to regulate micro-climates and groundwater, and can reduce peaks in intense rain runoff, reducing flash flooding downstream; reducing other stressors on ecosystems and habitat fragmentation; changing cropping, livestock, and aquaculture practices; and investing in green infrastructure.

Both the food and agriculture supply chain and the consumer products sectors are heavily dependent on biodiversity and ecosystems services, and so building resilience through investments in natural capital is particularly important to those two sectors. Natural capital could be enhanced in a range of agricultural commodities by

- Acting to avoid, reduce and reverse land degradation as this can increase food and water security and contribute to broader resilience.
- Diversifying water resources, enhancing watershed and reservoir management, and improving integrated water management. Adaptive water

management techniques include enhancing storage and access to irrigation water, more efficient water delivery systems, and agronomy that improves soil water retention.
- Improving tolerance of crops to high temperature, changing crop rotation systems, breeding additional drought-tolerant crop varieties, and maintaining genetic diversity.
- Soil conservation and restoration. Soil health is the continued capacity of soil to function as a vital, living ecosystem. This is achieved by cycling nutrients; absorbing, draining, and retaining rainwater; filtering water to remove pollutants; and increasing the diversity of soil animals and microorganisms.
- Reducing pesticides use. Practices that reduce the overall use of pesticides and steer farmers toward organic agriculture are favorable, as is reducing the cloud application of pesticides in favor of ground-level application in specific areas.
- Pest management. Insects are major limiting factors in producing cotton, and hundreds of species of insects may be found in cotton, some of which are important economically.[12]

Physical capital

Physical capital refers to infrastructure including those related to transport, manufacturing, real estate, logistics, and communications. It involves climate-proofing to ensure population centers are protected from climate extremes. Beyond infrastructure, physical capital refers to equipment, facilities, logistics, communications, utilities, and even genetic agricultural resources. There are numerous investments in physical capital that reduce risk, such as cyclone shelters, wind-resistant buildings, adopting and strengthening building codes, upgrading existing buildings, and improving utilities and transportation systems. Flood defenses are increasingly common in low-lying states and coastal regions. Better flood control is one low-hanging fruit as affordable and effective technologies already exist, including dams and dykes.[30] Investments in roads, bridges, and stronger protections for utility services are vital for ensuring continued links across supply chains and between workers, employers, and consumers in the aftermath of climate-related events. Early warning systems along with hazard, exposure, and vulnerability mapping provide companies with tools to aid disaster preparedness and enterprise risk management.

Vale is a Brazilian-based metal and mining company, ranking as one of the largest producers of iron ore and nickel in the world. The company invested more than $18.6 million to implement a short-term forecasting program at a weather monitoring center to issue weather warnings to prepare its port facility for extreme weather events. The data gathered from this early warning is used for forecasting and nowcasting (forecasts for the next 30 minutes to 3 hours), enabling Vale to closely monitor weather conditions; and, when shared with the local authorities, helps municipalities prepare for extreme weather events.[31]

These are affordable and highly cost-effective strategies. Adopting the latest building code requirements is affordable and saves $11 per $1 invested. Building 1 foot above the 100-year flood elevation is cost effective, adding only $90 million of construction cost per year for new construction while saving $550 million, a 6-to-1 benefit-cost ratio. Building-code hurricane requirements save an average of $10 per $1 of added cost ($5.6 billion saved for an annual investment of $540 million), with benefit-cost ratios that reach as high as 30 to 1. Activities that enhance resilience of water and wastewater facilities, electric utility substations, roads and railways, and communications equipment yielded benefit-cost ratios as high as 31 to 1.[32]

Financial capital

Financial capital refers to the volume of available financial resources and access to financial goods and services. Financial capital concerns both the mobilization of increased financial flows in support of resilience and the critical expansion of financial services to frontline sectors, companies, and communities. Companies can enhance financial capital through the provision of insurance schemes; income, asset, and livelihood diversification; the provision of catastrophe bonds; the development of microfinance products; and the facilitation of cash transfers to frontline communities. Improved employment practices can also enhance financial capital, as businesses can set the wage and working conditions for individuals impacted by climate change. Companies make strategic choices around whether to pay a living wage to their frontline workers, as well as whether these workers will receive the same benefits in insurance, paid leave, and predictable and flexible working hours that the white-collar workforce receives. This has implications for the financial resources individuals can call upon when faced with climate hazards.[33]

Weather index insurance has also become an important product to drive investments and safeguard against climate losses. In 2010, Sompo Japan Nipponkoa observed that farmers in northeast Thailand were suffering significant revenue losses as a result of extreme weather events and other climate impacts. Very few insurance products were available to protect them against these risks; and, as a result, extreme weather events – such as flooding and drought – posed serious threats to their ability to support themselves and their families. To address this problem, Sompo launched a new weather index insurance product, which also responded to increased demand for financial and insurance products to cover the revenue losses caused by extreme weather events. Weather index insurance products (weather derivatives) provide compensation and/or insurance payments to farmers when temperatures and rainfall breach certain thresholds or when other extreme weather events occur.[12]

Mobilizing financial capital can yield a resilience dividend. The National Institute of Building Sciences estimates investing in physical assets to support climate resilience saves up to $10 per $1 invested. For example, building-code hurricane requirements save an average of $5.6 billion for an annual investment of $540 million. The Benefit–Cost ratios can reach as high as 30 to 1 in areas along the Gulf of Mexico and Atlantic coastline due to their exacerbated risk of more intense and frequent hurricanes.[34] As both public and private investments shift away from expensive disaster response toward resilience; there are massive opportunities for companies who can build resilient infrastructure and communities.

Research commissioned by the Global Commission on Adaptation also found that investing $1.8 trillion globally in five areas from 2020 to 2030 could generate $7.1 trillion in total net benefits. The five areas the report considers are early warning systems, climate-resilient infrastructure, improved dryland agriculture, mangrove protection, and investments in making water resources more resilient.[35]

Political capital

Political capital refers to access to decision-making that shapes policy environments to enable resilience. Just as climate change undermines the realization of human rights, the strengthening of human rights is arguably the most important intervention to enhance resilience. Access to information helps vulnerable populations anticipate climate-related events and take preventative action. Being more considered in decision-making enables marginalized communities to help shape public policy in a manner that accounts for their specific vulnerabilities. And access to justice enables communities to hold both the public and private sector accountable for failures to build resilience in a manner that is proportional. Political capital is also critical to addressing the social, cultural, and economic inequalities that exacerbate risk of climate change. These inequalities include the differentiated vulnerability faced by women, indigenous peoples, and the urban poor. Companies can enhance political capital by advocating for reduced gender inequality and marginalization in all its forms; the extension of social safety nets and protection to women; and improved access to information, decision-making, justice, education, health, energy, and housing.

Because of the centrality of human rights to climate resilience, companies should adopt a policy clearly stating the corporate commitment to respect for human rights. This means that they should avoid infringing on the human rights of others and should address adverse human rights impacts with which they are involved. Companies should follow a three-step analysis to identify and clarify where they may have responsibilities to support the human rights of communities facing climate risks.

- Locate high-risk areas: All companies are connected to climate-vulnerable communities in some capacity, although often not in their direct operations. To determine the scope of responsibility and appropriate remedy, a company must first map its operations and the operations of significant business

partners to identify where the company operates in known high-risk climate-vulnerable communities.
- Clarify the scope of responsibility: Next, companies should determine whether they are "causing" or "contributing" to increasing the vulnerability of high-risk communities or whether they are directly linked to a business partner that is increasing the vulnerability of an at-risk community. Put differently, companies should ask whether their operations (or significant business partner) cause or contribute to weakening the realization of human rights that would protect vulnerable communities from the negative impacts of climate change.
- Determine the appropriate remedy. When a company is causing or contributing to an adverse human rights impact, it is obligated to cease or prevent that impact from occurring. Risk mitigation can take many forms. Where due diligence reveals increased vulnerability caused by a company's own operations, the remedy should include a focus on strengthening capital assets in those communities to provide a forward-looking remedy that strengthens the community's resilience to adverse climate events. The company may also need to consider changing certain aspects of its operations, including diversifying water sources, mitigating risk in the workforce, updating land-acquisition methods, and incorporating ways to build capacity to prevent gender discrimination in the workforce.[36]

Securing a just transition

The low-carbon and resilient economy cannot replicate the inequalities of today. Inequality manifests itself in a variety of forms. Intersecting economic, political, social, cultural, and legal inequalities are experienced by communities across the globe, with structural discrimination often based on income or class, race, ethnicity, religion, or nationality. The Human Development Report 2019, published by the United Nations Development Programme (UNDP), is one of the most comprehensive assessments of global inequality. Launched on 12 December 2019, weeks before the COVID-19 outbreak in Wuhan, the Report looks at inequality from multiple lenses. Its key findings reveal inequalities impacting health, life expectancy, education, social mobility, economic prospects, and even the transfer of inequality from one generation to the next. For example, 600 million people live in extreme income poverty, defined as earning less than $1.90 per day. When the UNDP considers issues beyond income-based poverty – such as deprivations in health, education, and standard of living – it concludes that about 1.3 billion people or 23% of the world's population lived in a state of "multidimensional poverty" between 2006 and 2017. The difference in life expectancy at birth between someone born in an industrialized country and someone born in a low human development country is 19 years. 42% of adults in low human development countries have a primary education, compared with 94% in industrialized countries. Only 3.2% of adults in low human development countries have a college or university level education compared with 29% in developed countries. Looking to the future, UNDP expects "business as usual" approaches to development and poverty alleviation to still result in 3 million children under five years old dying every year from preventable causes; while 225 million children are expected to be out of school.[37]

Gender inequality is a glaring example of how the world of December 2019 failed to deliver on the promise of universal human rights. Across the globe, women lack access to financial services including savings, insurance, credit, and investments and are restricted in their ownership of productive assets such as property, farms, and inheritance. Women are concentrated in informal and unprotected work. Nearly 90% of African women are not covered by labor relations laws; they have no minimum wage or social protection and no maternity benefits or old age pension.[38]

The low-carbon and resilient economy must do more than addressing the chemical composition of the atmosphere and improve enterprise risk management; it must provide for a "just transition" for those individuals, households, and communities currently invested in high-carbon development; and it must use this unique moment to create a shared prosperity.

The just transition is an economy-wide process that produces the plans, policies, and investments that build resilient economies and communities with green and decent jobs.[39] This requires job creation through seizing new economic opportunities while reducing the disruption people and communities face in the transition away from high-carbon business models. A just transition that achieves these objectives will generate economic vitality and stability.[40]

In 2015, the International Labour Organization (ILO) outlined some foundational principles to guide just transition processes. First, it is vital to establish strong social consensus on the goal and pathways to sustainability – in this case the creation of a low-carbon and resilient economy. This means engaging in a comprehensive social dialogue across social partners including government, trades unions, communities, and businesses. Second, public policies should be developed that secure rights at work. Third, there should be strong recognition of the gender dimension of socio-ecological challenges and the promotion of gender-transformative policies to promote equity. Third, joined-up government is essential to ensure that economic, environmental, social, education, training, and labor policies work in unison to create an effective enabling environment. And fourth, comprehensive analysis of the impacts on employment, followed by social protection programs to cover job losses and training programs to build skills for the new economy are vital.[41]

From the beginning, the just transition has focused on jobs. The ILO has predicted that employment will be affected in four ways as climate policies and corporate leadership define a low-carbon economy:

- The expansion of greener products, services, and infrastructure will translate into higher labor demand across many sectors of the economy and associated new jobs.

- Some existing jobs will be substituted as we shift from internal combustion engine manufacturing to electric vehicle production or from coal to solar power. This shift may be particularly difficult for low-skilled workers as they will need to retrain to compete for new jobs. Moreover, locations losing employment may not be the same places to benefit from the new jobs, and so geographic disparities may increase.
- Certain jobs or sectors may be eliminated – either phased out or massively reduced in numbers. This is particularly acute for fossil fuel producers and carbon-intensive industries or practices.
- Many existing trade jobs will be transformed and redefined as day-to-day workplace practices, skill sets, work methods, and job profiles are greened. For example, automobile workers will produce more electric cars, while farmers will engage in sustainable land management, address food loss and waste, and respond to changing dietary habits.[42]

The practicalities of this will be difficult. Each of the decarbonization pathways outlined above in food, fuel, and other sectors of the economy requires the deployment of new innovations and technologies; sunsetting old technologies and incumbent industries; and dislocation of communities, many of whom have already suffered from globalization and deindustrialization. Consider the case of India. India is utterly dependent on coal as it generates more than three-quarters of the country's electricity. Mining it and turning it into power accounts for a tenth of India's industrial production. Coal India, the world's largest state-owned coal mining company, employs over 370,000 people, and there are up to 500,000 working in the coal industry at large. Estimates are that between 10 million and 15 million people benefit indirectly from coal through social programs near mines. It is also a big source of revenue for state governments in poor areas.[43]

As we work to secure a prosperous future for communities sunsetting the high-carbon economy, we must avoid the trap of creating a low-carbon economy characterized by the same level of exploitation, inequality, and environmental harm. The new low-carbon economy will be mineral intensive because clean energy technologies need more materials than fossil-fuel-based electricity generation technologies. Analysis from the World Bank estimates that seventeen critical minerals and metals will be needed at scales significantly beyond current production levels. The demand for these minerals and metals is expected to rise substantially up to 2050, increasing in both absolute and percentage terms from 2018 production levels, with increases in demand of up to nearly 500% for certain minerals, especially those concentrated in energy storage technologies. For example, production of graphite, lithium, and cobalt will need to be significantly ramped up by more than 450 from 2018 levels by 2050 to meet demand from energy storage technologies. About 4.5 million tons of graphite will be needed annually by 2050, or a cumulative of 68 million tons. This represents a 500% increase from 2018 production figures. Copper is used across eight or more technologies. Deploying solar power at scale requires 20 million tons of copper by

2050, growth of approximately 350% compared with the base scenario. Twelve of the 17 minerals assessed by the World Bank show much larger increases in demand because of their use in energy storage. Graphite and nickel are both used in Li-ion batteries, the most widely projected deployed battery technology. Global demand for nickel is expected to jump 16× from 2018 levels, with storage technologies accounting for 73% of the demand.[44]

Many of these minerals will come from resource-rich developing countries and emerging economies – the very same communities and countries that consistently rank amongst the most vulnerable to climate impacts such as extreme weather events, flooding, drought, wildfires, water scarcity, and vector-borne diseases. Eight out of 11 countries with the largest copper reserves, and 8 out of the top 10 producers, are in emerging markets or developing countries, with Chile leading the way in both the reserves and production. Four of the top five countries with the highest nickel reserves are emerging countries. Indonesia and the Philippines are the top two producers of nickel.[45] The mining communities in these countries are already particularly vulnerable to climate change due to their direct reliance on the natural environment, often remote locations, sparse populations, and minimal infrastructure. An increase in climate-induced weather-related hazards such as forest fires, flooding, temperature extremes, and wind storms may further affect the viability of mining operations and undermine development pathways for these communities. Commercial trade in minerals and metals relies on road, rail, and sea transport. Sea-level rise may affect maritime transport infrastructure; while roads and rail may be adversely affected by heat stress, increased precipitation, flooding, and subsistence.[46] Climate change–induced health and safety effects include increased incidence of malaria and other vector-borne diseases. These impacts could decrease the resilience of host communities, risk weakening companies' social license to operate, and increase potential for conflict.

Mining operations in these communities may be perceived to exacerbate the extreme events. Today, 70% of mining operations from the six largest mining companies are located in water-stressed countries.[47] With climate change expected to reduce renewable surface water and groundwater resources significantly in most dry subtropical regions, competition for water among sectors is likely to increase with substantial implications for operations and relationships with host communities. Together, these climate risks could increase operating, transportation, and decommissioning costs; lead to a reduction or shut-down of production;[48] lead to falls in labor productivity; increase insurance rates and increase capital expenditure; exacerbate compliance and legal liability impacts;[49] and ultimately may undermine the ability of the sector to meet the growing demand for minerals and metals as the world drives toward net-zero emissions by 2050.

Ensuring that the rewards of building the new low-carbon and resilient economy of the 21st century are retained in and invested by the emerging markets, and developing countries who will fuel this transition becomes a critical new component of the just transition. These countries cannot be treated as communities from which to extract and then abandon. Companies can contribute to a just transition by working to strengthen social dialogue; social protection systems; vocational training; transfer payments; and investments in communities suffering deindustrialization and dislocation and by committing to creating shared value in the communities that will build our new 21st-century economy.

Conclusion

Companies must decarbonize to avoid unmanageable climate change. This means reducing greenhouse gas emissions by 45% below 2010 levels by 2030 and achieving net-zero emissions by 2050. The innovations needed to achieve these intermediate and long-term goals in food, energy, and transport systems are available to companies today. Food systems decarbonization requires a combination of land management, value chain management, risk management, and the proper use of enablers such as financial goods and services, public policy and consumer education that incentivize reforestation, reductions in food waste, changes in diets, and enhancement of soils as carbon sinks. Energy transformation involves increasing energy efficiency and moving from fossil fuels to renewable energy such as wind and solar. It further involves modernizing our understanding of energy transmission with investments in modern grids, more local generation through micro-grids, and investments in storage technology. Transformation in transport is focused on modal shifts and a switch from the internal combustion engine to electric vehicles. It also involves rethinking how cities are designed and reimagining consumption patterns and global supply chains to enable more goods and services to be produced and consumed locally.

Companies must also manage unavoidable climate change by building socio-ecological resilience. This means investing in the six so-called capital assets – human, social, natural, physical, financial, and political – that are the key building blocks of resilience. It means supporting governments in the creation of social protection programs, mobilizing finance at scale to drive decarbonization, expanding access to finance to build adaptive capacity, and strengthening the human rights of vulnerable populations.

Finally, companies must work to secure a just transition for all, ensuring that the new climate economy is one of shared prosperity both for those communities currently invested in a high-carbon path and for those who will build the low-carbon future.

References

1. Bradshaw, C.J.A., Ehrlich, P.R., Beattie, A., Ceballos, G., Crist, E., Diamond, J., Dirzo, R., Ehrlich, A.H., Harte, J., Harte, M.E., Pyke, G., Raven, P.H., Ripple, W.J., Saltré, F., Turnbull, C., Wackernagel, M. and Blumstein, D.T. (2021) "Underestimating the Challenges of Avoiding a Ghastly Future". *Frontiers in Conservation Science* 1:615419. https://www.frontiersin.org/articles/10.3389/fcosc.2020.615419/full

2. Ritchie, H. (2020) "Sector by Sector: Where Do Global Greenhouse Gas Emissions Come from?" In: *Our World in Data*. Published on 18 September, 2020. https://ourworldindata.org/ghg-emissions-by-sector
3. Christensen, J. and Olhoff, A. (2019) *Lessons from a Decade of Emissions Gap Assessments*. Nairobi: United Nations Environment Programme. https://wedocs.unep.org/bitstream/handle/20.500.11822/30022/EGR10.pdf
4. Levin, K., Rich, D., Ross, K., Fransen, T. and Elliott, C. (2020) *Designing and Communicating Net-Zero Targets*. Washington, DC: World Resources Institute. https://www.wri.org/publication/designing-and-communicating-net-zero-targets
5. IPCC (Intergovernmental Panel on Climate Change) (2019) *Climate Change and Land: An IPCC special report on climate change, desertification, land degradation, sustainable land management, food security, and greenhouse gas fluxes in terrestrial ecosystems* [P.R. Shukla, J. Skea, E. Calvo Buendia, V. Masson-Delmotte, H.-O. Pörtner, D. C. Roberts, P. Zhai, R. Slade, S. Connors, R. van Diemen, M. Ferrat, E. Haughey, S. Luz, S. Neogi, M. Pathak, J. Petzold, J. Portugal Pereira, P. Vyas, E. Huntley, K. Kissick, M. Belkacemi, J. Malley, (eds.)]. Geneva: IPCC. https://www.ipcc.ch/site/assets/uploads/2019/08/Fullreport-1.pdf
6. Jia, G., Shevliakova, E., Artaxo, P., De Noblet-Ducoudré, N., Houghton, R., House, J., Kitajima, K., Lennard, C., Popp, A., Sirin, A., Sukumar, R. and Verchot L. (2019) "Land–Climate Interactions". In: *Climate Change and Land: An IPCC Special Report on Climate Change, Desertification, Land Degradation, Sustainable Land Management, Food Security, and Greenhouse Gas Fluxes in Terrestrial Ecosystems*. Geneva: Intergovernmental Panel on Climate Change (IPCC). https://www.ipcc.ch/site/assets/uploads/2019/08/2c.-Chapter-2_FINAL.pdf
7. Arneth, A., Denton, F., Agus, F., Elbehri, A., Erb, K., Osman Elasha, B., Rahimi, M., Rounsevell, M., Spence, A. and Valentini R. (2019) "Framing and Context". In: *Climate Change and Land: An IPCC Special Report on Climate Change, Desertification, Land Degradation, Sustainable Land Management, Food Security, and Greenhouse Gas Fluxes in Terrestrial Ecosystems*. Geneva: Intergovernmental Panel on Climate Change (IPCC). https://www.ipcc.ch/site/assets/uploads/2019/08/2b.-Chapter-1_FINAL.pdf
8. Jia, G., Shevliakova, E., Artaxo, P., De Noblet-Ducoudré, N., Houghton, R., House, J., K. Kitajima, C. Lennard, A. Popp, A. Sirin, R. Sukumar, L. Verchot (2019) "Land–Climate Interactions". In: *Climate Change and Land: An IPCC Special Report on Climate Change, Desertification, Land Degradation, Sustainable Land Management, Food Security, and Greenhouse Gas Fluxes in Terrestrial Ecosystems*. Geneva: Intergovernmental Panel on Climate Change (IPCC). https://www.ipcc.ch/site/assets/uploads/2019/08/2c.-Chapter-2_FINAL.pdf
9. Jia, G., E. Shevliakova, P. Artaxo, N. De Noblet-Ducoudré, R. Houghton, J. House, K. Kitajima, C. Lennard, A. Popp, A. Sirin, R. Sukumar, L. Verchot (2019) "Land–climate interactions". In: *Climate Change and Land: An IPCC special report on climate change, desertification, land degradation, sustainable land management, food security, and greenhouse gas fluxes in terrestrial ecosystems*. Geneva: Intergovernmental Panel on Climate Change (IPCC). https://www.ipcc.ch/site/assets/uploads/2019/08/2c.-Chapter-2_FINAL.pdf
10. Pearce, A. and Berkenkamp, J. (2017) *Assessing Corporate Performance on Food Waste Reduction: A Strategic Guide for Investors*. Washington, DC: Natural Resources Defense Council (NRDC). https://www.nrdc.org/sites/default/files/corporate-performance-food-waste-reduction-ib.pdf
11. Food and Agriculture Organization of the United Nations (FAO). (2013) *Food Wastage Footprint Impacts on Natural Resources: Summary Report*. Rome: FAO. http://www.fao.org/3/i3347e/i3347e.pdf
12. Pieper, M., Michalke, A. and Gaugler, T. (2020) "Calculation of External Climate Costs for Food Highlights Inadequate Pricing of Animal Products". *Nature Communications 11*: 6117. https://www.nature.com/articles/s41467-020-19474-6

13. IPCC (Intergovernmental Panel on Climate Change) (2019) "Summary for Policymakers". In: *Climate Change and Land: An IPCC Special Report on Climate Change, Desertification, Land Degradation, Sustainable Land Management, Food Security, and Greenhouse Gas Fluxes in Terrestrial Ecosystems*. Geneva: IPCC. https://www.ipcc.ch/site/assets/uploads/2019/08/Edited-SPM_Approved_Microsite_FINAL.pdf
14. Cameron, E. (2019) *Business Adaptation to Climate Change and Global Supply Chains*. The Hague: The Global Commission on Adaptation. https://gca.org/reports/business-adaptation-to-climate-change-and-global-supply-chains/
15. Jacobson, M.Z. Delucchi, M.A., Bauer, Z., Goodman, S., Chapman, W., Cameron, M., Bozonnat, C., Chobadi, L., Clonts, H., Enevoldsen, P., Erwin, J., Fobi, S., Goldstrom, O., Hennessy, E., Liu, J., Lo, J., Meyer, C., Morris, S., Yachanin, A. (2017) "100% Clean and Renewable Wind, Water, and Sunlight All-Sector Energy Roadmaps for 139 Countries of the World". *Joule 1(1)*: 108–121, doi:10.1016/j.joule.2017.07.005. https://www.sciencedirect.com/science/article/pii/S2542435117300120
16. Hawken, P. (ed.) (2017) *Drawdown: The Most Comprehensive Plan Ever Proposed to Reverse Global Warming*. London: Penguin Books.
17. Levin, K. (2018) *New Global CO_2 Emissions Numbers Are In. They're Not Good*. Washington, DC: World Resources Institute. https://www.wri.org/blog/2018/12/new-global-co2-emissions-numbers-are-they-re-not-good
18. Hawken, P. (ed.) (2017) *Drawdown: The Most Comprehensive Plan Ever Proposed to Reverse Global Warming*. London: Penguin Books.
19. IPCC (Intergovernmental Panel on Climate Change) (2018) "Global Warming of 1.5°C. An IPCC Special Report on the Impacts of Global Warming of 1.5°C above Pre-Industrial Levels and Related Global Greenhouse Gas Emission Pathways". In: *The Context of Strengthening the Global Response to the Threat of Climate Change, Sustainable Development, and Efforts to Eradicate Poverty*. Geneva: IPCC. https://www.ipcc.ch/site/assets/uploads/sites/2/2019/06/SR15_Full_Report_Low_Res.pdf
20. Lawrence Livermore National Laboratory. (2018) *Americans Ramp Up Use of Solar, Wind Energy*. Livermore, CA: Lawrence Livermore National Laboratory. https://www.llnl.gov/news/americans-ramp-use-solar-wind-energy
21. Mathers, J. (2020) *Walmart Commits to 100% Zero-Emission Trucks by 2040, Signaling Electric Is the Future*. Published on 22 September, 2020. Washington, DC: Environmental Defense Fund (EDF). http://blogs.edf.org/energyexchange/2020/09/22/walmart-commits-to-100-zero-emission-trucks-by-2040-signaling-electric-is-the-future/
22. Hawkins, A. (2020) "Amazon Unveils Its New Electric Delivery Vans Built by Rivian". In: *The Verge*. Published on 8 October, 2020. New York: Vox Media. https://www.theverge.com/2020/10/8/21507495/amazon-electric-delivery-van-rivian-date-specs
23. More information on the GLEC Framework (2021) is available from The Smart Freight Centre at: https://www.smartfreightcentre.org/en/how-to-implement-items/what-is-glec-framework/58/
24. Mulligan, J., Ellison, G., Levin, K., Lebling, K. and Rudee, A. (2020) *6 Ways to Remove Carbon Pollution from the Sky*. Washington, DC: The World Resources Institute. https://www.wri.org/insights/6-ways-remove-carbon-pollution-sky
25. IPCC (Intergovernmental Panel on Climate Change). (2012) *"Summary for Policymakers"*. In: *Managing the Risks of Extreme Events and Disasters to Advance Climate Change Adaptation: A Special Report of Working Groups I and II of the Intergovernmental Panel on Climate Change*. Cambridge, UK and New York: Cambridge University Press. https://www.ipcc.ch/site/assets/uploads/2018/03/SREX_Full_Report-1.pdf

26. Cameron, E., Harris, S. and Prattico, E. (2018) *Resilient Business, Resilient World: A Research Framework for Private-Sector Leadership on Climate Adaptation*. San Francisco: BSR. https://www.bsr.org/reports/BSR_Resilient_Business_Resilient_World_A_Research_Framework_for_Private_Sector_Leadership_on_Climate_Adaptation.pdf
27. Keating, A., Campbell, K., Mechler, R., Michel-Kerjan, E., Mochizuki, J., Kunreuther, H., et al. (2014) *Operationalizing Resilience against Natural Disaster Risk: Opportunities, Barriers and a Way Forward*. Zurich: Zurich Flood Resilience Alliance. http://opim.wharton.upenn.edu/risk/library/ZAlliance-Operationalizing-Reslience.pdf
28. Cameron, E., Lemos, M. and Winterberg, S. (2018) *Climate and Inclusive Economy: The Business Case for Action*. San Francisco: BSR. https://www.bsr.org/en/our-insights/report-view/climate-inclusive-economy-the-business-case-for-action
29. Costanza, R., Pérez-Maqueo, O., Martinez, M.L., Sutton, P., Anderson, S.J. and Mulder, K. (2008) "The Value of Coastal Wetlands for Hurricane Protection". *AMBIO: A Journal of the Human Environment* 37(4): 241–248. Stockholm: Royal Swedish Academy of Sciences. https://www.sciencedirect.com/science/article/pii/S2212041612000101
30. United Nations Office for Disaster Risk Reduction (UNDRR). (2020) *The Human Cost of Disasters: An Overview of the Last 20 Years (2000–2019)*. Geneva: UNDRR. https://www.undrr.org/news/drrday-un-report-charts-huge-rise-climate-disasters
31. United Nations Global Compact. (2015) *The Business Case for Responsible Corporate Adaptation: Strengthening Private Sector and Community Resilience – A Caring for Climate Report*. New York: UN global Compact. https://www.unglobalcompact.org/library/3701
32. Porter, K., Dash, N., Huyck, C., Santos, J., Scawthorn, C., Eguchi, M., Eguchi, R., Ghosh, S., Isteita, M., Mickey, K., Rashed, T., Reeder, A., Schneider, P. and Yuan, J. (2019) *Natural Hazard Mitigation Saves: 2019 Report*. Washington, DC: Multi-Hazard Mitigation Council/National Institute of Building Sciences. https://www.nibs.org/files/pdfs/NIBS_MMC_MitigationSaves_2019.pdf
33. Cameron, E., Lemos, M. and Winterberg, S. (2018) *Climate and Inclusive Economy: The Business Case for Action*. San Francisco: BSR. https://www.bsr.org/en/our-insights/report-view/climate-inclusive-economy-the-business-case-for-action
34. Porter, K., et al. (2019) *Natural Hazard Mitigation Saves: 2019 Report*. Washington, DC: Multi-Hazard Mitigation Council at the National Institute of Building Sciences. https://www.nibs.org/files/pdfs/NIBS_MMC_MitigationSaves_2019.pdf
35. Global Commission on Adaptation (GCA). (2019) *Adapt Now: A Global Call for Leadership on Climate Resilience*. The Hague: GCA. https://gca.org/reports/adapt-now-a-global-call-for-leadership-on-climate-resilience/
36. Cameron, E. and Nestor, P. (2018) "Climate and Human Rights: The Business Case for Action". In: *BSR Report*. San Francisco: BSR. https://www.bsr.org/en/our-insights/report-view/climate-human-rights-the-business-case-for-action
37. United Nations Development Programme (UNDP). (2019) "Beyond Income, beyond Averages, beyond Today: Inequalities in Human Development in the 21st Century". In: *Human Development Report 2019*. New York: UNDP. http://hdr.undp.org/sites/default/files/hdr2019.pdf
38. Aguilar, L., Granat, M. and Owren, C. (2015) *Roots for the Future: The Landscape and Way Forward on Gender and Climate Change*. Washington, DC: IUCN & GGCA. https://wedo.org/wp-content/uploads/2015/12/Roots-for-the-future-final-1.pdf
39. B Team. (2017) *Just Transition: A Business Guide*. New York: B Team. https://bteam.org/our-thinking/reports/just-transition-a-business-guide
40. Wei, D. (2018) *Climate and the Just Transition: The Business Case for Action*. San Francisco: BSR. https://www.bsr.org/en/our-insights/report-view/climate-just-transition-the-business-case-for-action

41. ILO (International Labor Organization) (2015) *Guidelines for a Just Transition towards Environmentally Sustainable Economies and Societies for All*. Geneva: ILO. https://www.ilo.org/wcmsp5/groups/public/—ed_emp/—emp_ent/documents/publication/wcms_432859.pdf
42. ILO (International Labor Organization) (2016) *A Just Transition to Climate-Resilient Economies and Societies: Issues and Perspectives for the World of Work*. Geneva: ILO. https://www.ilo.org/wcmsp5/groups/public/—ed_emp/—gjp/documents/publication/wcms_536552.pdf
43. The Economist. (2018) "The Black Hole of Coal: India Shows How Hard It Is to Move beyond Fossil Fuels". *The Economist 432: 9162*. The Economist. London.
44. Hund, K., La Porta, D., Fabregas, T.P., Laing, T. and Drexhage, J. (2020) *Minerals for Climate Action: The Mineral Intensity of the Clean Energy Transition*. Washington, DC: The World Bank Group. https://pubdocs.worldbank.org/en/961711588875536384/Minerals-for-Climate-Action-The-Mineral-Intensity-of-the-Clean-Energy-Transition.pdf
45. Hund, K., La Porta, D., Fabregas, T.P., Laing, T. and Drexhage, J. (2020) *Minerals for Climate Action: The Mineral Intensity of the Clean Energy Transition*. Washington, DC: The World Bank Group. https://pubdocs.worldbank.org/en/961711588875536384/Minerals-for-Climate-Action-The-Mineral-Intensity-of-the-Clean-Energy-Transition.pdf
46. Bourgouin, F. (2014) *Climate Change: Implications for Extractive and Primary Industries. Key findings from the Intergovernmental Panel on Climate Change Fifth Assessment Report*. Cambridge: University of Cambridge Judge Business School/Cambridge Institute for Sustainability Leadership. https://www.bsr.org/reports/BSR-Cambridge-Climate-Change-Implications-for-Extractives.pdf
47. International Finance Corporation and International Council on Mining and Minerals. (2017) *Shared Water, Shared Responsibility, Shared Approach: Water in the Mining Sector*. Washington, DC, and London: IFC and ICMM. https://www.icmm.com/en-gb/research/environmental-stewardship/shared-water-shared-responsibility
48. Mason, L. and Giurco, D. (2013) *Climate Change Adaptation for Australian Minerals Industry Professionals*. Gold Coast: National Climate Change Adaptation Research Facility. https://www.researchgate.net/publication/323357353_Climate_change_adaptation_for_Australian_minerals_industry_professionals_Final_Project_Report
49. Golder. (2021) *Guide on Climate Change Adaptation for the Mining Sector*. Ottawa: The Mining Association of Canada. https://mining.ca/wp-content/uploads/2021/05/MAC-Climate-Change-Guide-April-30.pdf

5

CURRENT CORPORATE LEADERSHIP

How commitments and inconsistencies shape the private sector response to climate crisis

Over 6,000 companies and investors, representing $36 trillion in revenue, have committed to climate action. These companies, from 120 countries, make up half the global economy.[1] These pioneering companies are compelling their suppliers to disclose climate risk and set emissions reductions as a business condition.[2] Across every facet of the economy, companies are pledging to only procure renewable energy, draw the era of the internal combustion engine to a close, end commodity-driven deforestation, and change the way they invest and spend their money. The financial sector is beginning to move as well. Financial institutions with cumulative assets of at least $47 trillion under management (representing 25% of global financial market) have set climate-related targets for their portfolios.[3]

These companies are embracing the role of first movers as they understand the need to avoid the physical risks of climate change and minimize the transition risks as the global economy accelerates the shift to a low-carbon economy. They are also embracing a series of emerging opportunities:

- First, there is the opportunity to capture new and substantial investments. Collectively, the national climate plans governments have developed under the Paris Agreement represent at least a $13.5 trillion market for the energy sector alone. This includes $8.3 trillion in improvements in energy efficiency in the transport, buildings, and industry sectors alone.[4] When other sectors such as transport, agriculture, industrial process, and cities are included the number is considerably higher. In the past year, the G20 have mobilized more than $14.9 trillion in economic recovery, most of which is required to be climate compatible.[5] And in the United States alone, the Biden Administration is pursuing climate and infrastructure spending in excess of $2 – the equivalent of 4.5% of Federal spending on climate and infrastructure spending.⁶ The signals are clear – substantial sums of public money will be spent in the coming years on laying the foundations for achieving the net-zero goal. Companies in the vanguard of climate leadership are seeking a piece of that financial pie.

- Second, there is also an opportunity to capture a generational shift in investment preferences. The amount of assets under management (AUM) for sustainable investments has grown to $8.72 trillion in the US, a 33% increase since 2014 and now constitutes 22% of the $40.3 trillion of total AUM in the United States. Millennials are driving this change – putting money in sustainable investments at a rate 2× higher than average. Eighty-six percent of millennial investors say they are "very interested" or "interested" in sustainability; 61% have made at least one sustainable investment in the last year; and 75% think their investments can influence climate change. With a $30 trillion wealth transfer coming to millennials over the coming decades, this is a powerful trend that leading corporations are keen to benefit from.⁷
- Third, there are the opportunities of the so-called "resilience dividend." The National Institute of Building Sciences estimates investing in physical assets to support climate resilience saves up to $10 per $1 invested. For example, building-code hurricane requirements save an average of $5.6 billion for an annual investment of $540 million. The Benefit-Cost-Ratios can reach as high as 30 to 1 in areas along the Gulf of Mexico and Atlantic coastline due to their exacerbated risk of more intense and frequent hurricanes.⁸ As earlier chapters have illustrated, avoiding the costs of disrupted supply chains, minimizing operational shortfalls, and sidestepping reputational damage also form part of the resilience dividend.

However, this corporate climate leadership is not all that it may at first appear to be. Commitments are plagued by inconsistencies. While many companies have real plans to back up their pledges, others are being accused of setting rhetorical commitments, devoid of real investment, and lacking in implementation strategies. Corporate commitments are also undermined by a key contradiction, namely the problems that occur when an aspiration to decarbonize collides with the business instinct to pursue ever-increasing levels of growth. These inconsistencies and contradictions invite reputational harm on individual companies, undermine the overall integrity of corporate climate commitments, and jeopardize efforts to avoid unmanageable climate change through decarbonization.

It should also be stressed that the emergence of corporate climate leadership remains embryonic. While six thousand companies seems like an impressive figure, it still only represents approximately 10% of the 60,000 multinational corporations.⁹ Moreover, of the 1,693 multinational corporations listed as committed to climate action by We Mean Business; more than 60% are headquartered in Europe and the United States, suggesting that corporate climate leadership remains rooted in the global north.¹⁰

This chapter explores the current climate leadership and assesses its many strengths and increasingly problematic weaknesses. The first section examines a range of climate commitments including science-based targets; net-zero goals; renewable energy targets; and efforts to build resilience. The second section looks at sector-wide climate ambition including efforts to transform transport, food systems, and finance. The third section explores the inconsistencies and contradictions and explains why the increasing gap between rhetoric and reality risks undermining the progress that has been made in creating this unprecedented level of corporate climate leadership.

Corporate climate commitments

The explosion of corporate climate commitments over the past decade covers every sector of the global economy. The vast majority of commitments focus on greenhouse gas emissions reductions, and of these, the largest concentration is on energy systems. The We Mean Business coalition has been to the fore in driving corporate uptake of climate commitments and now lists eleven different commitment areas as part of a portfolio of pledges. These include commitments to adopt so-called science-based emissions reductions targets; commitments to renewable energy and energy productivity; the deployment of electric vehicle fleets; climate-smart agriculture; and carbon pricing. This section explores some of the commitments major corporations are now taking, revealing their potential to accelerate the transition to a low-carbon economy.

Science-based targets (SBTs) and the net-zero goal

Science-based targets have long been considered the benchmark for corporate climate leadership. Corporate targets are considered "science-based" if they are in line with what the latest climate science deems necessary to meet the goals of the Paris Agreement – limiting global warming to well below 2°C above pre-industrial levels and pursuing efforts to limit warming to 1.5°C.

Increasingly companies are aligning their science-based targets with the goal of holding global mean temperature rises below 1.5°C above pre-industrial levels. This means reaching 45% emissions reductions by 2030 and net-zero emissions by 2050. More than 1,300 companies are currently committed to science-based Targets.[11] The two key steps in setting and achieving these targets are first to quantify the company's greenhouse gas baseline; and second to develop tailored strategies to reduce absolute emissions across the full value chain predominantly by addressing energy, transport, land use, and manufacturing emissions.

If Science-based Targets were considered the benchmark in recent years, they are increasingly being replaced by net-zero targets, the new gold standard in corporate climate leadership. The term "net-zero emissions" means achieving a balance between reducing greenhouse gas pollution caused by human activities at source and removal of residual emissions using sinks. A source is understood as any unit or process that releases greenhouse gasses into the atmosphere (e.g. fuel combustion, energy generation, livestock management, etc.), and a sink includes natural systems such as trees, soils, and oceans that are capable of depositing greenhouse gasses into long-term storage.

Greenhouse gases (GHGs) have different life spans in the atmosphere and different global warming potentials. Carbon dioxide (CO_2) and Nitrous Oxide (N_2O) are "long-lived" GHGs, meaning they exist in the atmosphere for hundreds of years and accumulate. "Short-lived" GHGs, such as Methane (CH_4) and most HFCs, exist in the atmosphere for a shorter period (e.g. 12 years in the case of Methane) but typically have higher global warming potential than CO_2. Most countries have set net-zero targets that cover all six GHGs because it sends strong political signals and is better aligned with the Paris Agreement, which requires a balance between sources and sinks of GHGs in the second half of the century. Consequently, corporate net-zero targets should also cover all GHGs and not just CO_2.[12]

This complex equation translates into simple guidance for companies – GHG emissions released to the atmosphere from sources within the company's value chain should not exceed GHGs removed from the atmosphere by sinks resourced and sustained by the company. This balance between sources and sinks should cover the full basket of six greenhouse gasses and should be achieved by 2050.

At least one-fifth (21%) of the world's 2,000 largest public companies, with sales of nearly $14 trillion, have committed to meet net-zero targets.[13] The growth in these commitments is even more impressive than the baseline. The NewClimate Institute recently reported a threefold increase in the number of businesses setting net-zero goals, from 500 at the end of 2019 to 1,565 in October 2020, with even companies in emissions-intensive industries, such as fossil fuels, materials, and transportation services, setting ambitious targets.[14]

Embarking on a net-zero target can be complex as there are many competing definitions and criteria governing the commitments. The Race-to-Zero campaign has created so-called "starting line criteria" that provide a useful roadmap for how companies can embark on a net-zero journey. The four steps are:

1. A leadership pledge from the corporation's senior leadership to reach net-zero emissions by mid-century at the latest.
2. A tangible plan that explains what steps will be taken toward achieving net-zero with short- and medium-term goals, including an intermediate target that identifies emissions reductions by 2030.
3. Immediate and verifiable action toward achieving net-zero, consistent with delivering the 2030 interim target.
4. A commitment to public disclosure of progress at least annually.[15]

Companies are also guided to cover all emissions, including full value chain emissions and not just those from their own activities, and from using offsets sparingly and at a decreasing rate until they are phased out entirely by the time net-zero is achieved.

Agriculture and land use commitments

The Intergovernmental Panel on Climate Change (IPCC) has concluded that up to 30% of total food produced is lost or wasted. Reducing food waste by 50% would generate net emissions as high as 30% of total food-sourced GHGs, equivalent to more than 10% of annual emissions and slightly more than the GHG emissions of the entire European Union.[16] 32 of the world's 50 largest food companies by revenue have responded to this by creating and joining programs that have set a food loss and waste reduction target. For example, ten of the world's leading food retailers and providers have engaged 200 major suppliers to reduce food loss and waste through the 10 × 20 × 30 initiative. These companies are committing to measure and report food waste in their own operations; act to reduce food waste both in their own operations and in partnership with suppliers; create collaborations to look at systemic issues; and educate customers to enable after-sales reduction in food waste.[17]

Reaching net-zero by 2050 requires a balance between sources and sinks. Forests offer an opportunity to address both. According to the IPCC, reduced deforestation is one of the three most important measures to reduces land use greenhouse gases at source, while afforestation and reforestation are among the most significant contributors to enhancing sinks through Carbon Dioxide Removal (CDR).[18] Seventeen multinational consumer goods firms with a collective market value of $1.8 trillion have responded to this by committing to tackle deforestation, forest degradation, and land conversion in supply chains through the Forest Positive Coalition of Action.[19] The coalition is working to accelerate efforts to remove commodity-driven deforestation from their individual supply chains; to set higher expectations for suppliers and traders to act across their entire supply chains; to drive transformational change in key commodity landscapes including palm oil, soy, paper, pulp, fiber, and beef; and to define measurable outcomes on which all companies agree to track and report.[20]

General Mills,[21] Cargill,[22] and Walmart[23] have each committed to regenerative agriculture. The term regenerative agriculture is often used to describe practices aimed at promoting soil health by restoring soil's organic carbon. The world's soils store several times more amount of carbon than that in the atmosphere, acting as a natural "carbon sink." The IPCC estimates the mitigation potential for soil organic matter stocks is equivalent to almost 20% of global GHGs.[18] Practices grouped under regenerative agriculture include no-till agriculture – where farmers avoid plowing soils and instead drill seeds into the soil – and use of cover crops, which are plants grown to cover the soil after farmers harvest the main crop. Other practices include diverse crop rotations, such as planting three or more crops in rotation over several years and rotating crops with livestock grazing.[24]

Energy and transport commitments

The majority of corporate commitments cover the energy sector and focus on increasing the amount of renewable energy as a share of the energy system; rapidly transitioning corporate energy use away from fossil fuels; increasing energy productivity, particularly in high-carbon sectors; and enabling a shift toward the use of electric vehicles.

Renewable Energy (RE) 100 is one of the better-known energy commitments. A global initiative bringing together almost 300 companies across 125 markets and collective revenue of over $2.75 trillion. The aggregate energy demand of all companies participating in RE100 exceeds that of Australia. The fundamentals are that participating companies pledge to procure all their energy needs from renewable sources. To achieve 100% renewable electricity, companies must match 100% of the electricity used across their global operations with electricity produced from renewable sources. These can include biomass (including biogas), geothermal, solar, water, and/or wind – either sourced from the market or self-produced. RE100 companies must select a target date for achieving 100% renewable electricity, with the minimum requirements being 100% by 2050, 60% by 2030, and 90% by 2040.[25]

Changing the energy system requires sending powerful signals to energy utilities in order to encourage them to shift from fossil fuels, invest in energy infrastructure, and arrive at a price point that makes clean energy profitable for them and affordable for corporate consumers. The Renewable Energy Buyers Alliance (REBA) is an important coalition working on this market signal. Starting in 2014, REBA has become an alliance of large clean energy buyers, developers, and service providers and today counts more than 200 large energy buyers and over 150 clean energy developers and service providers among its members. Participants in the REBA community have been a part of 95% of all large-scale US corporate renewable energy deals to date; and accounted for 97% of the renewable energy deals in 2020, with Amazon, Google, Verizon, McDonald's, Facebook, and General Motors constituting the top six renewable energy purchasers by volume.[26] These deals create new incentives and consequences for the major utilities.

The transport sector is also undergoing significant change, perhaps even a paradigm shift. Over ten carmakers have committed to EV sales targets for the period between 2020 and 2025. Volkswagen will invest $66 billion by 2024 to build a 40% EV fleet by 2030.[27] Toyota plans to generate half of its sales from electrified vehicles by 2025. Ford plans to have 40 EV models by 2022. Volvo has committed to generate 50% of its global sales from EVs by 2025 and reduce the total carbon footprint of each vehicle manufactured by 40%. Fiat Chrysler plans to offer more than 30 EV models by 2022.[28] The shift is not limited to car companies but also extends into the owners of large vehicle fleets. Walmart has announced that it will replace its conventional vehicle fleet with zero-emission vehicles by 2040.[29]

Decarbonization of financial flows

Financial institutions have a significant role to play in accelerating the transition to a low-carbon economy, and in recent years, companies in the financial services sector have begun to shift their investments toward decarbonization.

According to Morgan Stanley, the amount of assets under management (AUM) for sustainable investments has grown to $8.72 trillion in the United States, a 33% increase since 2014 and now constitutes 22% of the $40.3 trillion of total AUM.[7] Globally, financial institutions with cumulative assets of at least $47 trillion under management (representing 25% of global financial market) have set climate-related targets for their portfolios.[30] These range from setting specific investment targets (e.g. Goldman Sachs pledged $750 billion for sustainable finance by 2030)[31] to phasing out coal (e.g. BNP Paribas[32] and Crédit Agricole[33]) and net-zero commitments (e.g. Barclays[34] and HSBC[35] announced plans to get to net zero by 2050, not just in their own operations but also for emissions that they finance).

The Royal Bank of Scotland (RBS) has developed a policy that could become the gold standard. In February 2020, RBS announced that it would at least halve the climate impact of its financing activity by 2030 and will end financing for major oil and gas companies unless they have credible transition plans in place by the end of 2021, to align with the Paris Agreement.[36] In September 2020, Morgan Stanley announced a new commitment to reach net-zero financed emissions by 2050. They pledged to take a leadership role in developing the tools and methodologies needed to measure and manage GHG emissions across the financial services sector.[37]

Blackrock is the world's largest asset manager with nearly $7 trillion in investments. In 2017, the CEO announced plans to make environmental sustainability a core goal. They own more oil, gas, and thermal coal reserves than any other investor, with total reserves amounting to 30% of total energy-related emissions from 2017. They are currently the world's largest investor in coal plant developers, holding shares worth $11 billion. They plan to: begin to exit investments that "present a high sustainability-related risk," such as those in coal; Introduce new funds that shun fossil fuel-oriented stocks; vote against management teams that are not making progress on sustainability.[38] In 2020, Blackrock identified 244 companies that are making insufficient progress in integrating climate risk into their business models or disclosures. Of these companies, they took voting action against 53 or 22%. They have put the remaining 191 companies "on watch." Those that do not make significant progress risk voting action against management in 2021. Voting action means they voted against the re-election of one or more members of a company's board, voted against the discharge of directors or the entire board, or voted for one or more climate-related shareholder proposals. In 2020, Blackrock voted against a number of Exxon's directors for their failure to have long-term greenhouse gas reduction targets; and the company's lack of disclosure around the degree of warming it expects under its stated strategy.[39]

In addition to shifting finance away from carbon-intensive and toward low-emission development, financial institutions are also encouraging greater private sector transparency on exposure to climate risk through disclosure reporting. Financial Institutions with a combined $150 trillion in assets under management, and 1,500 non-financial corporations with a combined market capitalization of $13 trillion, have signed up to the Task Force on Climate-related Financial Disclosures (TCFD) committing to improve and increase reporting of climate-related financial risk.[40]

Resilience commitments

Much of the focus for corporate climate leadership has been on greenhouse gas emissions reductions. Indeed, one of the core arguments of this book is that a more balanced view of climate leadership, encompassing resilience and inclusion, is needed during the coming decade. Despite the relative paucity of resilience commitment areas, there are some important initiatives that provide a foundation upon which to build.

Many companies are now turning to the Global Organic Textile Standard (GOTS) and Textile Exchange's Organic Content Standard (OCS) to guarantee farm-level chain of custody and ensure that the cotton they source and sell is organic, free of harmful chemicals, and uses these environmentally sustainable practices to preserve and restore natural capital. GOTS is recognized as the world's leading processing standard for textiles made from organic fibers as it defines high-level environmental criteria covering the processing, manufacturing, packaging, labeling, trading, and distribution of all textiles made from at least 70% certified organic natural fibers. Williams–Sonoma, Nike, Patagonia, and Adidas are among the leading multinationals turning to this certification system. Williams–Sonoma is working toward a goal of 100% responsibly sourced cotton by 2021 using GOTS as a standard.[41]

One leading proponent of human capital advancement in the cotton sector is the BCI – the largest cotton sustainability program in the world. Established by WWF (formerly the World Wildlife Fund) and supported by leading companies such as H&M (formerly Hennes & Mauritz), Adidas, and Ikea, BCI aims to train five million farmers, covering 30% of global cotton production, in sustainable farming practices by 2020. In 2016–2017, Better Cotton was grown in 21 countries by 1.3 million licensed BCI farmers and accounted for 14% of global cotton production. BCI works through Implementing Partners (IPs) – including non-governmental organizations (NGOs) and companies – to help farmers acquire the social and environmental knowledge they need to cultivate Better Cotton. Each IP supports more than 4,000 Field Facilitators who, in turn, run Learning Groups across communities and regions to master best practice techniques in line with the Better Cotton Principles and Criteria, which define sustainable cotton through seven key standards, including focusing attention on enhancing adaptive capacity and reducing greenhouse gas emissions from land use associated with cotton production.[42]

Inconsistencies and contradictions

This corporate climate leadership is not all that it may at first appear to be. Commitments are plagued by inconsistencies. While many companies have real plans to back up their pledges, others are being accused of setting rhetorical commitments, devoid of real investment, and lacking in implementation strategies, and according to the NewClimate Institute only 8% of corporate net-zero targets include interim targets to chart a decarbonization pathway.[14] In other words, they have made commitments that come due 30 years into the future when all of the company's leaders are long-retired and safe from any accountability. Another group of companies are pursuing contradictory corporate policies – committing to decarbonization within their own operations while continuing to channel large sums to the fossil fuel sector. Some companies are embracing climate ambition through their sustainability strategies while actively undermining climate decarbonization and resilience building through their single-minded pursuit of profit and their public policy advocacy. This section examines some of the inconsistencies and contradictions and argues that they invite reputational harm on individual companies, undermine the overall integrity of corporate climate commitments, and jeopardize efforts to avoid unmanageable climate change through decarbonization.

Inconsistent commitments leading to accusations of greenwashing

The Climate Ambition Alliance brings together countries, businesses, investors, cities, and regions that are working toward achieving net-zero CO_2 emissions by 2050. According to the Alliance, more than 1,600 corporations have committed to net-zero emissions. This is an impressive number, given the concept barely existed at the time of the Paris climate agreement. However, there is a difference between a net-zero commitment and a robust plan to achieve it. There is a gathering consensus that a robust commitment to net-zero should include coverage of all emissions including Scope 3 (emissions that are the result of activities from assets not owned or controlled by the company but that the company indirectly impacts in its value chain); and that offsetting will meet robust and will reduce over time to the minimum feasible. When this definition of net-zero is used, only 11 companies clear the bar.[43] This increasingly leaves companies open to accusations of corporate greenwashing and a sense that they are being performative in their commitments rather than outlining practical strategies.

Many companies choose to ignore their Scope 3 emissions from net-zero goals. The Greenhouse Gas Protocol separates emissions into three scopes. It's mandatory for businesses to report on both Scope 1 (direct emissions) and Scope 2 (indirect emissions from the energy they buy). But most companies are not required to report on Scope 3 emissions, which include indirect emissions from the supply chain, transportation of products, consumer use, product disposal, etc. Companies' Scope 3 emissions are on average 11.4 times higher than their operational emissions.⁴⁴ As a result, the failure to account for these emissions completely undermines the integrity of the targets.

Some companies have been stung from previous criticism and spurred on to greater action. Apple was an early adopter of the RE100 target. However, in the infancy of its commitment it claimed to be a design and retail company and so not responsible for the full value chain emissions of its products and services. Consequently, it applied a 1,000% renewable energy target to only 4% of its actual Scopes 1, 2, and 3 emissions. That results in a 4% target rather than a 100% one. The criticism was swift and loud. Apple has learned from its lesson and when announcing a net-zero goal in 2020 was clear that the Plan would be applied across its entire business, manufacturing supply chain, and product life cycle.⁴⁵

There is an old saying that a goal without a plan is just a wish. At first glance, corporate climate commitments look like ambitious roadmaps for decarbonization, but at closer inspection most of them fall into the wishes category as they lack the clarity, interim milestones, and tangible steps that would make them both operational and verifiable. Many companies have committed to a 2050 target without offering any clarity on what they will achieve during this decade nor on how they plan to achieve the target. Only 8% of companies with net-zero goals have interim targets.¹⁴ Some with a 2050 target are claiming the reputational reward for announcing the target while then continuing with business as usual or worse. For example, in 2019 Amazon launched the "Climate Pledge" to great fanfare, and yet their emissions went up by 15% in 2020 as they pursued a high growth strategy.⁴⁶ Moreover, Amazon considers itself a technology company rather than a retailer and does not include Scope 3 emissions in its own calculations. The fossil fuel companies are perhaps the most audacious in this regard, BP and Shell have both committed to net-zero by 2050, but their growth plans involve extracting and burning 120% more fossil fuels than the limit for keeping the planet under 1.5°C of warming.⁴⁷

Corporate net-zero targets are built by combining direct reduction of emissions with offsets.¹⁴ Offsets are any activity that compensates for the emission of greenhouse gases by providing for an emission reduction elsewhere. This can include funding renewable energy projects that replace coal-fired power plants or investing in carbon sequestration in soils or forests. This is a legitimate approach to the sources and sinks balance that is inherent to the net-zero concept, but only if used sparingly. Many companies are making commitments to net-zero without disclosing how much of the emissions reductions will be met through offsetting. In essence, many seem to be working with the assumption that they can ignore their emissions as long as they contribute to afforestation programs to enhance carbon sinks. This approach could work if a limited number of companies relied on offsets, but the more companies follow this approach without actual emission cuts, the more difficult it becomes to achieve global net-zero emissions.⁴⁸

A recent Client Earth study looking at FTSE 250 companies' annual reports concluded that when it came to net-zero targets, meaningful detail about assumptions, methodologies, and strategies regarding targets was often limited or missing.⁴⁹ As Baker has suggested, companies must be honest: How much carbon pollution will still be taking place? How much will they be relying on removal? How exactly will they be reducing emissions? Will they be making actual changes to their business model?⁴⁷ The integrity of the whole corporate climate leadership enterprise rests on the answers to these questions.

Contradictions with business goals undermining climate leadership

If the inconsistencies are undermining the integrity of climate commitments, the contradictions are undermining the hopes of achieving decarbonization by mid-century. While there are many contradictions, this section focuses on the preoccupation with growth. With financial institutions, this manifests as an obsession with returns and results in ongoing investments in fossil fuel companies, even by banks making climate commitments. In other parts of the private sector, this is illustrated by a fixation on meeting growth targets and driving consumption.

There is an old adage in the sustainable development arena that we will only manage what we measure, and we only measure what we value. The real economy values growth above all else, and this is one of the principal drivers of the climate crisis. Our efficiency gains are simply insufficient to keep pace with the rate of resources we consume and the amount of greenhouse gas pollution emitted in the production, consumption, and disposal of consumer goods and services.

Dauvergne has written about a rebound effect, whereby the gains resulting from companies making environmental commitments are greatly outweighed by their continued commitment to growth, particularly by driving consumption.⁵⁰

For example, car manufacturers sold more than 87 million new vehicles in 2015, 14 million more than in 2010. At the current pace, the world is heading toward as many as 3 billion cars on its roads by 2050.⁵⁰ The International Organization of Motor Vehicle Manufacturers (OICA) estimates that there are over 1 billion passenger cars in the world today. China was the world's third-largest car market in 2006. Today, 1 out of 3 cars produced in the world comes from China, and with vehicle penetration in China at only about 150 vehicles per 1,000 people, compared with approximately 700 vehicles per 1,000 people in the mature markets of the G7, that number is set to rise.⁵¹ Cars and car parts exported by China are responsible for nine times more CO_2 per dollar than those exported by Germany.⁵² Even with the switch to entirely electric vehicle fleets the rise in absolute numbers of private passenger vehicles will likely strain efforts to decarbonize if produced and used in the wrong locations. On average, about half of the lifetime emissions from an electric vehicle comes from making the battery. A medium-sized battery made in renewables rich Sweden admits around 350 kg of CO_2. For coal-reliant Poland, that figure is over 8 tons.⁵² Given these trends, can automakers' commitment to electric vehicle fleets compensate for their desire to sell more cars?

The same questions apply to other sectors. Unilever, the poster-child for corporate climate leadership, recently announced it would be "laser focused" on driving top-line sales growth to achieve a long-term underlying sales growth target of 3% to 5%. A growth rate of 5% would lead to a doubling of the company within 15 years. Unilever's main rival in the consumer products sector is Proctor & Gamble (P&G). It has announced a consumer sales growth forecast to a range of 5% to 6%.[53] Both companies have science-based targets and renewable energy targets. Unilever also has targets on electric vehicles, carbon pricing, and climate-smart agriculture. Walmart, while claiming to be among the world's most sustainable companies, is guided by an entire corporate philosophy that is about growth and expansion, promoting consumption of more low-priced, short-term, non-durable products. Indeed, since its 2005 sustainability pledge, Walmart has added 3,000 new big-box stores worldwide to the 1,600 it had at the time.[54] It may be possible for some companies to sustain this level of growth while decarbonizing, but there is little evidence to suggest that the level of decoupling of economic progress from socio-ecological resources could be achieved at global scale.

The issue is not just the obsession with growth; the source of that growth is also important. Chapter 1 of this book highlighted how complex global supply chains amplified climate risk. Those same supply chains also increase emissions. China's exports alone now account for about 5% of the world's fossil fuel emissions. Most of this relates to goods that are ultimately consumed in the developed world as two-thirds of China's emissions exports go to members of the OECD. Moreover, according to figures from the British government, the carbon emissions caused by transporting a given weight by air are about 70 times greater than if it had been shipped.[52] This means sectors reliant on timely delivery, such as fast fashion, consumer products, and food, are particularly damaging to the climate system.

The financial services sector is vital to accelerating the transition to a low-carbon economy. The International Energy Agency estimates that globally it could take $3.5 trillion in energy sector investments alone every year through 2050 to reorient toward a climate-neutral economy.[55] However, financial institutions are currently pursuing contradictory strategies that undermine decarbonization in their relentless pursuit of ever-higher returns.

The world's biggest 60 banks have provided $3.8 trillion of financing for fossil fuel companies since the Paris climate Agreement in 2015. Finance to 100 of the biggest expanders of coal, oil, and gas fell by 20% between 2016 and 2018, but last year bounced back 40% in 2019. At this rate, fossil financing will hit $1 trillion per year by 2030 and undermine efforts to achieve net-zero by 2050.[36] Consider the contradictions in the investment patterns of the following banks:

- JPMorgan Chase has committed to procure 100% of its energy needs from renewable sources and yet is responsible for $269 billion in fossil fuel financing between 2016 and 2019.
- Wells Fargo has committed to procure 100% of its energy needs from renewable sources and has invested $198 billion in fossil fuels between 2016 and 2019.

- Citi has committed to procure 100% of its energy needs from renewable sources and has invested $188 billion in fossil fuels between 2016 and 2019.
- Bank of America has made commitments to procure all of its energy needs from renewable sources and to have an exclusively electric vehicle fleet. However, the Bank pumped $157 billion into fossil fuels since 2016 and was the largest non-Chinese coal power funder in 2019.
- Barclays has committed to procure 100% of its energy needs from renewable sources and has invested $118 billion in fossil fuels between 2016 and 2019.
- TB Bank has committed to procure 100% of its energy needs from renewable sources and has invested $103 billion in fossil fuels between 2016 and 2019.

Each of these banks needs to recognize that the discrepancies between their commitments and their investment portfolios are ticking time bombs, not just for the integrity of their own brands, but the wider integrity of corporate climate ambition.

A number of prominent economists and philosophers, including Amartya Sen and Kate Raworth, have proposed ending our fixation on growth. Sen argues that economic development should be seen as a process of expanding capabilities through the deepening of political freedoms, economic facilities, social opportunities, transparency guarantees, and protective security. This not only involves expanding political and civil rights but also improving social arrangements such as education, healthcare, and access to financial services. Sen argues that a narrow focus on growth prioritizes economic well-being rather than the well-being of individuals and the societies in which they live.[56] Raworth points out that, today, we have economies that need to grow, whether or not they make us thrive; when, in fact, we need economies that make us thrive, whether or not they grow.[57] Writing for the WEF, Jennifer Morgan, Executive Director of Greenpeace International, has proposed a well-being index informed by the UN's Sustainable Development Goals.[58] Instead of assessing our well-being based on what we produce and consume, we would instead have metrics related to poverty alleviation; food security; quality and access to education; gender equality; access to and quality of water, sanitation and energy; our ability to innovate and build 21st-century infrastructure; our progress in providing decent work with a living wage for our population; the sustainability of our cities and the viability of our circular economy; our efforts to manage the climate crisis; and our ability to safeguard biodiversity on land and in the oceans. These prescriptions, when translated into corporate strategies, would in effect provide a means to measure and therefore value decarbonization and the enhancement of socio-ecological resilience.

Conclusion

This chapter presented the unprecedented scale and speed of corporate climate commitments. The emergence of this corporate climate leadership over the past decade, and in particular since the adoption of the Paris Agreement, has been one of the most positive and optimistic developments in the fight against global climate change. However, as the chapter has revealed, these commitments are inconsistent, and there continue to be significant contradictions within the private sector that undermine the ultimate goal of building a low-carbon and resilient world. Over the coming years a variety of measures need to be taken to build on the leadership foundations that have emerged:

First, there needs to be a dramatic increase in the volume and geographical diversity of corporate climate commitments. The goal of net-zero emissions by 2050 cannot be achieved so long as 90% of multinational corporations sit on the sidelines, particularly when so many of those companies are in emerging markets and the global south.

Second, companies must work to improve the integrity of corporate pledges. The new corporate climate leadership will need robust plans rather than rhetorical commitments to achieve net-zero.

And finally, companies need to go confront the contradictions at the heart of their climate commitments.

References

1. Hsu, A., Widerberg, O., Weinfurter, A., Chan, S., Roelfsema, M., Lütkehermöller, K. and Bakhtiari, F. (2018) "Bridging the Emissions Gap – The Role of Non-State and Subnational Actors". In: *The Emissions Gap Report 2018. A UN Environment Synthesis Report*. Nairobi: United Nations Environment Programme. https://wedocs.unep.org/bitstream/handle/20.500.11822/26093/NonState_Emissions_Gap.pdf?sequence=1
2. A sample of commitments taken by 1600 companies can be accessed through "We Mean Business". https://www.wemeanbusinesscoalition.org
3. More information available from UNEP FI: https://www.unepfi.org/net-zero-alliance/
4. Wei, D., Cameron, E., Harris, S., Prattico, E., Scheerder, G. and Zhou, J. (2016) *The Paris Agreement: What It Means for Business*. New York: We Mean Business. https://www.bsr.org/en/our-insights/report-view/the-paris-agreement-on-climate-what-it-means-for-business
5. Harvery, F. (2021) "Global Green Recovery Plans Fail to Match 2008 Stimulus, Report Shows". In: *The Guardian*, 12 February, 2021. Manchester: The Guardian. https://www.theguardian.com/environment/2021/feb/12/global-green-recovery-plans-fail-to-match-2008-stimulus-report-shows
6. The White House. (2021) "The American Jobs Plan". *White House Fact Sheet* issued on 31 March, 2021. https://www.whitehouse.gov/briefing-room/statements-releases/2021/03/31/fact-sheet-the-american-jobs-plan/
7. Morgan Stanley Institute for Sustainable Investing. (2017) *Sustainable Signals: New Data from the Individual Investor*. New York: Morgan Stanley & Co. LLC. https://www.morganstanley.com/pub/content/dam/msdotcom/ideas/sustainable-signals/pdf/Sustainable_Signals_Whitepaper.pdf
8. Porter, K. et al. (2019) *Natural Hazard Mitigation Saves: 2019 Report*. Washington, DC: Multi-Hazard Mitigation Council at the National Institute of Building Sciences. https://www.nibs.org/files/pdfs/NIBS_MMC_MitigationSaves_2019.pdf
9. SciencePo. (2018) "Multinational Corporations ". In: *World Atlas of Global Issues, 2018*. Published on 28 September, 2018. Paris: SciencePo. https://espace-mondial-atlas.sciencespo.fr/en/topic-strategies-of-transnational-actors/article-3A11-EN-multinational-corporations.html

10. Information accessed from www.wemeanbusiness.org on 11 April, 2021.
11. The number of companies adopting science-based targets is tracked and validated by the Science-based Targets Initiative (SBTi). https://sciencebasedtargets.org/companies-taking-action
12. Levin, K., Rich, D., Ross, K., Fransen, T. and Elliott. C. (2020) "Designing and Communicating Net-Zero Targets". *WRI Working Paper*. Washington, DC: World Resources Institute (WRI). www.wri.org/design-net-zero.
13. Shetty, D. (2021) "A Fifth of World's Largest Companies Committed to Net Zero Target". In: *Forbes*, 24 March, 2021. New Jersey: Forbes Media. https://www.forbes.com/sites/dishashetty/2021/03/24/a-fifth-of-worlds-largest-companies-committed-to-net-zero-target/?sh=4208cc88662f
14. Day, T., Mooldijk, S., Kuramochi, T., Hsu, A., Yi Yeo, Z., Weinfurter, A., Xi Tan Y., French, I., Namdeo, V., Tan, O., Raghavan, S., Lim, E. and Nair, A. (2020) *Navigating the Nuances of Net-Zero Targets*. Berlin: NewClimate Institute & Data-Driven EnviroLab. https://newclimate.org/wp-content/uploads/2020/10/NewClimate_NetZeroReport_October2020.pdf
15. The Race to Zero "starting line" criteria. https://unfccc.int/sites/default/files/resource/Minimum-criteria-for-participation-in-RTZ.pdf
16. Jia, G., Shevliakova E., Artaxo P., De Noblet-Ducoudré N., Houghton R., House J., Kitajima K., Lennard C., Popp A., Sirin A., Sukumar R. and Verchot L. (2019) "Land–Climate Interactions". In: *Climate Change and Land: An IPCC Special Report on Climate Change, Desertification, Land Degradation, Sustainable Land Management, Food Security, and Greenhouse Gas Fluxes in Terrestrial Ecosystems*. Geneva: Intergovernmental Panel on Climate Change (IPCC). https://www.ipcc.ch/site/assets/uploads/2019/08/2c.-Chapter-2_FINAL.pdf
17. Systemiq. (2020) *The Paris Effect: How the Climate Agreement Is Reshaping the Global Economy*. London: Systemiq. https://www.systemiq.earth/wp-content/uploads/2020/12/The-Paris-Effect_Full-Report-1.pdf
18. IPCC. (2019) "Summary for Policymakers". In: *Climate Change and Land: An IPCC Special Report on Climate Change, Desertification, Land Degradation, Sustainable Land Management, Food Security, and Greenhouse Gas Fluxes in Terrestrial Ecosystems*. Geneva: Intergovernmental Panel on Climate Change (IPCC). https://www.ipcc.ch/site/assets/uploads/2019/08/Edited-SPM_Approved_Microsite_FINAL.pdf
19. IU Energy. (n.d.) "Consumer goods giants worth $1.8trn team up to tackle deforestation". https://iuenergy.co.uk/consumer-goods-giants-worth-1-8trn-team-up-to-tackle-deforestation/
20. The Consumer Goods Forum. (2021) *Taking Root: Embarking on the Forest Positive Journey: Reflections and Ambitions from the Consumer Goods Forum (CGF) Forest Positive Coalition of Action*. Paris: The Consumer Goods Forum. https://www.theconsumergoodsforum.com/wp-content/uploads/2021/03/CGF-FPC-Taking-Root-Embarking-on-the-Forest-Positive-Journey-2021.pdf
21. General Mills. (2020) *Regenerative Agriculture 2020*. Golden Valley: General Mills. https://www.generalmills.com/en/Responsibility/Sustainability/Regenerative-agriculture
22. Cargill. (2020) *Cargill to Advance Regenerative Agriculture Practices across 10 Million Acres of North American Farmland by 2030*. Minnetonka: Cargill. https://www.cargill.com/2020/cargill-to-advance-regenerative-agriculture-practices-across-10
23. McMillon, D. (2020) *Walmart's Regenerative Approach: Going Beyond Sustainability*. Bentonville: Walmart. https://corporate.walmart.com/newsroom/2020/09/21/walmarts-regenerative-approach-going-beyond-sustainability

24. Ranganathan, J., Waite, R., Searchinger, T. and Zionts, J. (2020) "Regenerative Agriculture: Good for Soil Health, but Limited Potential to Mitigate Climate Change" *WRI Blog*. Published on 12 May, 2020. Washington, DC: World Resources Institute. https://www.wri.org/blog/2020/05/regenerative-agriculture-climate-change
25. The Climate Group. (2020) *Growing Renewable Power: Companies Seizing Leadership Opportunities*. London: The Climate Group. https://www.there100.org/growing-renewable-power-companies-seizing-leadership-opportunities
26. Renewable Energy Buyers Alliance (REBA). (2021) "REBA Announces Top 10 U.S. Large Energy Buyers in 2020". *Press Release*, 10 February, 2021. Washington, DC: REBA. https://rebuyers.org/blog/reba-announces-top-10-u-s-large-energy-buyers-in-2020-2/
27. Rauwald, C. (2019) "VW Challenges Rivals with $66 Billion for Electric Car Era". In: *Bloomberg*. https://www.bloomberg.com/news/articles/2019-11-15/vw-boosts-new-technology-spending-to-66-billion-through-2024
28. Levin, T. (2020) "All the Things Carmakers Say They'll Accomplish with Their Future Electric Vehicles between Now and 2030". In: *Business Insider*, 28 January, 2020. New York: Insider Inc. https://www.businessinsider.com/promises-carmakers-have-made-about-their-future-electric-vehicles-2020-1
29. Kane, M. (2020) "Walmart to Electrify Entire Fleet by 2040". In: *InsideEVs*, 24 September, 2020. https://insideevs.com/news/445649/walmart-electrify-entire-fleet-by-2040/?fbclid=IwAR18HR7x7BzNTtRi0BVz14VZiPf2yESh43b4GcUKu6vgca0uVkJE_1Awcds
30. Lütkehermöller, K., Mooldijk, S., Roelfsema, M., Höhne, N. and Kuramochi, T. (2020) *Unpacking the Finance Sector's Investment Commitments*. Berlin: NewClimate Institute and Utrecht University. https://newclimate.org/wp-content/uploads/2020/09/NewClimate_Unpacking_Finance_Sector_Sept20.pdf
31. Goldman Sachs. (n.d.) *Sustainable Finance*. New York: Goldman Sachs. https://www.goldmansachs.com/our-commitments/sustainability/sustainable-finance/
32. BNP Paribas. (2020) *BNP Paribas Is Accelerating Its Timeframe for a Complete Coal Exit*. Paris: BNP Paribas. https://group.bnpparibas/en/press-release/bnp-paribas-accelerating-timeframe-complete-coal-exit
33. Institute for Energy Economics and Financial Analysis (IEEFA). (2019) *France's Crédit Agricole to Stop Thermal Coal Investments in EU, OECD by 2030*. Paris: IEEFA. https://ieefa.org/frances-credit-agricole-to-stop-thermal-coal-investments-in-eu-oecd-by-2030/
34. Barclays. (2020) *Our Ambition to Be a Net Zero Bank by 2050*. London: Barclays Bank. https://home.barclays/society/our-position-on-climate-change/.
35. HSBC. (2020) *HSBC Sets Out Net Zero Ambition*. London: HSBC. https://www.hsbc.com/news-and-media/hsbc-news/hsbc-sets-out-net-zero-ambition
36. Rainforest Action Network (RAN). (2020) *Banking on Climate Change: Fossil Fuel Financing Report 2020*. San Francisco: RAN. http://priceofoil.org/content/uploads/2020/03/Banking_on_Climate_Change_2020.pdf
37. Stanley M. (2020) "Morgan Stanley Announces Commitment to Reach Net-Zero Financed Emissions by 2050". *Press Release*. New York: Morgan Stanley. https://www.morganstanley.com/press-releases/morgan-stanley-announces-commitment-to-reach-net-zero-financed-e
38. Fink, L. (2019) "A Fundamental Reshaping of Finance". *Annual Letter to Shareholders*. New York: Blackrock. https://www.blackrock.com/corporate/investor-relations/larry-fink-ceo-letter
39. Blackrock. (2017) *Our Approach to Sustainability*. New York: Blackrock. https://www.blackrock.com/corporate/literature/publication/our-commitment-to-sustainability-full-report.pdf?utm_source=newsletter&utm_medium=email&utm_campaign=newsletter_axiosgenerate&stream=top

40. TCFD (2020) "Task Force on Climate-related Financial Disclosures (2020)". *Status Report*. New York: TCFD. https://www.fsb.org/wp-content/uploads/P291020-1.pdf
41. Williams–Sonoma. (2017) *Corporate Responsibility Scorecard 2017*. San Francisco: Williams–Sonoma Inc. https://ir.williams-sonomainc.com/investor-information/news-releases/news-release-details/2018/Williams-Sonoma-Inc-Releases-2017-Corporate-Responsibility-Scorecard-Demonstrating-Leadership-in-the-Home-Industry/default.aspx
42. Better Cotton Initiative. (2017) *Annual Report*. Geneva: BCI. https://2017.bciannualreport.org
43. Black, R., Cullen, K., Fay, B., Hale, T., Lang, J., Mahmood, S. and Smith, S.M. (2021) Taking Stock: A Global Assessment of Net Zero Targets, Energy & Climate Intelligence Unit and Oxford Net Zero. https://ca1-eci.edcdn.com/reports/ECIU-Oxford_Taking_Stock.pdf?mtime=20210323005817&focal=none
44. CDP. (2021) *Transparency to Transformation: CDP Global Supply Chain Report 2021*. London: CDP. https://6fefcbb86e61af1b2fc4-c70d8ead6ced550b4d987d7c03fcdd1d.ssl.cf3.rackcdn.com/cms/reports/documents/000/005/554/original/CDP_SC_Report_2020.pdf?1612785470
45. Apple Inc. (2020) "Apple Commits to Be 100 Percent Carbon Neutral for Its Supply Chain and Products by 2030". *Press Release*, 21 July, 2020. Cupertino: Apple Inc. https://www.apple.com/newsroom/2020/07/apple-commits-to-be-100-percent-carbon-neutral-for-its-supply-chain-and-products-by-2030/
46. Arcieri, K. and Tsao, S. (2020) "Amazon's Emissions Increase 15% in 2019 amid Efforts to Reduce Carbon Footprint". In: *S&P Market Intelligence*, 5 August, 2020. London: S&P. https://www.spglobal.com/marketintelligence/en/news-insights/latest-news-headlines/amazon-s-emissions-increase-15-in-2019-amid-efforts-to-reduce-carbon-footprint-59261693
47. Baker, J. (2021) "How Not to Commit to 'Net Zero': 5 Common Carbon Strategy Mistakes". In: *Forbes, 3 March, 2021*. New York: Forbes. https://www.forbes.com/sites/jessibaker/2021/03/03/how-not-to-commit-to-net-zero-5-common-carbon-strategy-mistakes/?sh=5f09290b6bae
48. Black, R., Cullen, K., Fay, B., Hale, T., Lang, J., Mahmood, S. and Smith, S.M. (2021) *Taking Stock: A Global Assessment of Net Zero Targets*. Oxford: Energy & Climate Intelligence Unit and Oxford Net Zero. https://ca1-eci.edcdn.com/reports/ECIU-Oxford_Taking_Stock.pdf?mtime=20210323005817&focal=none
49. ClientEarth. (2021) *Accountability Emergency: A Review of UK-Listed Companies' Climate Change-Related Reporting (2019–20)*. London: ClientEarth. https://www.clientearth.org/media/wbglw3r3/clientearth-accountability-emergency.pdf
50. Dauvergne. (2016) *Environmentalism of the Rich*. Cambridge, MA: The MIT Press.
51. Statistics taken from the International Organization of Motor Vehicle Manufacturers (OICA) website on 11 April, 2021. https://www.oica.net
52. The Economist. (2019) "Trade and Emissions: Out of Sight". *The Economist 433(9165)*: 72–73. 19th October, 2019. London: The Economist. https://www.economist.com/finance-and-economics/2019/10/17/greta-thunberg-accuses-rich-countries-of-creative-carbon-accounting
53. Cavale, S. (2021) "Unilever's Back to the Future Goals Disappoint". In: *Reuters*, 21 February, 2021. https://www.reuters.com/article/uk-unilever-results/unilevers-back-to-the-future-goals-disappoint-idUSKBN2A40RV
54. Bakan, J. (2020) *The New Corporation: How "Good" Corporations Are Bad for Democracy*. New York: Vintage Books.

55. Herren Lee, A. (2020) "Big Business's Undisclosed Climate Crisis Plans". In: *The New York Times*, 27 September, 2020. New York: The New York Times Company. https://www.nytimes.com/2020/09/27/opinion/climate-change-us-companies.html?referringSource=articleShare
56. Sen, A. (1999) *Development as Freedom*. New York: Alfred A. Knopf.
57. Raworth, K. (2017) *Doughnut Economics: Seven Ways to Think Like a 21st-Century Economist*. White River Junction: Chelsea Green Publishing.
58. Morgan, J. (2020) *COVID-19 Is an Unmissable Chance to Put People and the Planet First*. Geneva: World Economic Forum (WEF). https://www.weforum.org/agenda/2020/04/covid19-coronavirus-climate-jennifer-morgan-greenpeace/

PART III
The new corporate climate leadership

6
THE INNOVATION AGENDA

Reimagining the climate-compatible business

Writing in Rolling Stone Magazine in 2017, the noted climate scientist Bill Mckibbon stated that on climate change, "winning Slowly Is the Same as losing."[1] The reason is clear. As Chapter 1 illustrated, the extent and implications of the climate crisis have been known for decades. Each decade of delay increases the costs of inaction, elevates the costs of action, and creates a need for greater urgency. At a certain point, delay eliminates the value of incremental progress and leaves transformation as the only available course of action. We have reached that moment now at the start of this new decade.

The United Nations Environment Programme (UNEP) has described the 2010s as "a decade lost" in a recent assessment of climate policies. Describing the findings as "sobering," the analysis reveals that global greenhouse gas (GHG) emissions in 2018 were almost exactly at the level of emissions projected for 2020 under the "business as usual" or "no policy" scenarios used in the Emissions Gap Report of 2011. In other words, there has been no real change in the global emissions pathway over the last decade.[2] This is remarkable given that 140 countries endorsed the Copenhagen Accord in 2009, with 85 of them pledging to reduce their emissions through national policies. The unprecedented Paris Agreement of 2015 was subsequently adopted by 195 countries, 184 of which have so-called nationally determined contributions designed to limit global warming to well below 2°C. The analysis goes on to report that because GHG emissions continue to grow, governments must now triple the level of ambition reflected in their current and planned climate policies to get on track toward limiting warming to below 2°C, while at least a fivefold increase is needed to align global climate action and emissions with limiting warming to 1.5°C by the end of this century.[2] Going forward, overcoming the challenges created by the lost decade will now require additional annual energy-related investments of between $830 billion and $2.4 trillion – about 2.5% of the world GDP.[3]

The private sector has made great strides in recent years to understand and respond to the climate crisis. Companies around the world and across industrial sectors have committed to reducing greenhouse gas emissions, advocated for ambitious climate policies, and worked to resource supply chain decarbonization and resilience. However, current levels of ambition remain insufficient and inconsistent, and so corporate leadership needs to be dynamic.

This chapter presents a framework for injecting new dynamism into corporate climate leadership through a three-point framework called act, enable, and accountability.

- *Act*: What should the new corporate climate leader do to improve its own understanding of climate risks, reduce its own greenhouse gas emissions, and contribute to building socioecological resilience? This section will offer a three-dimensional diagnosis of climate risk; provide a roadmap for improving corporate decarbonization commitments; and present the design elements of a corporate commitment to resilience.
- *Enable*: How does the new corporate climate leader enable decarbonization and resilience in others? This section provides guidance on how companies can enable greater climate ambition by addressing the inequality that amplifies climate risk, empowering communities' disproportionality impacted by climate change, and engaging in the process of debanking, reinvesting, and scaling finance to support the transition to a new climate economy.
- *Accountability*: How does the new corporate climate leader influence the emerging policy enabling the environment to create a global market conducive to decarbonization and resilience? How does a new corporate climate leader overcome the inconsistencies and contradictions that undermine decarbonization and resilience? How can the new corporate climate leader advance transparency to ensure common information and effort across the whole economy? This section will explore how companies should work to influence the creation of a policy enabling environment catalytic of business leadership on climate; overcome the current inconsistencies and contradictions in corporate climate leadership; and encourage companies to scale up disclosure to provide transparency and accountability to the market.

In the immediate aftermath of the adoption of the Paris Agreement, the authors designed a leadership framework called "act, enable, and influence" to drive future corporate climate leadership. This framework has been used over the past six years in sustainability consulting across a range of organizations and with dozens of companies. The authors have chosen to update the framework as:

- First, while the term "influence" was always intended to reflect efforts to shape a progressive public policy agenda on climate change, it can be too

narrowly understood as corporate lobbying, which is often understood as companies pursuing rent-seeking or self-interest rather than the common good.
- Second, the term "accountability" is more expansive and accounts for a necessary evolution. Current corporate commitments, while welcome, are often lacking in robustness, often deliberately obscure, and therefore lack integrity. As businesses seek to hold public policy-makers accountable for the creation of a policy-enabling environment, they should also be held to a higher level of accountability for their own commitments. And finally, disclosing climate risk and response strategies is increasingly viewed as the transparency vital to improving the functioning of the market in an era of physical and transition risk.

Act – driving decarbonization and resilience inside the new corporate climate leader

(Properly) diagnosing climate risk

Some companies are beginning to address climate risks by building on existing business risk assessment activities and integrating adaptation initiatives into enterprise-wide risk management systems.[4] However, most businesses are misdiagnosing climate risk and failing to build comprehensive strategies for resilience. Research conducted by BSR and CDP revealed that 72% of suppliers accept that climate risks could significantly impact their business operations, revenue, or expenditure, yet only half of these are currently managing this risk.[5] A separate assessment reviewed more than 1,600 corporate adaptation strategies and found significant blind spots in companies' analysis of climate risks. More than half of reporting companies expect that climate change will increase their operational costs and/or reduce or disrupt production capacity; however, these companies appeared to be misunderstanding how climate change can manifest in business impacts, from lost consumer purchasing power to employee absenteeism to raw material shortages.[6] What are these companies doing wrong?

Most companies are aware of the existence of climate-related hazards, such as extreme weather events, and the likely exposure of their resources and operations to the increase in frequency and intensity of these events. However, a growing body of research suggests that the majority of companies are blind to the third dimension – vulnerability – the specific underlying weaknesses that increase susceptibility to harm.[7] Vulnerability is the propensity of exposed elements, whether people, ecosystems, biodiversity, economic sectors, complex supply chains, or individual companies, to suffer adverse effects when climate-related hazards occur. It is the underlying weaknesses that exacerbate risk. This can be a weakness in infrastructure – Verizon suffering billions of dollars in damages because their copper wiring disintegrated in saltwater; or it can be a weakness in the workforce – failing to understand how and why intersecting social, cultural, economic, economic, and political inequalities exacerbate risk for certain populations. Consider the following two company scenarios:

- Company A is headquartered in Amsterdam, the largest cocoa port, the largest importer of cocoa beans, and home to the largest cocoa grinding industry in the world. The company is part of cluster, generating over €2.3 billion in annual revenues for the Dutch economy in a global sector worth $130 billion. Clients include the largest confectionary companies in the world such as Mars, Nestlé and Mondelez International, and Ferrero. The company is heavily dependent on cocoa from Cote d'Ivoire and Ghana, which together account for 70% of global cocoa production. Company A probably knows that erratic rainfall, flooding, and soil erosion are expected to lead to decreased production in Ghana and Cote d'Ivoire. However, the company is likely blind to the social, political, and economic conditions in these countries that amplify vulnerability. Women do about 70% of the work on Ivory Coast cacao farms but only receive only about 20% and own less than 25% of the land. They live way below the poverty line on less than $0.30 per day.
- Company B is a US-Dutch conglomerate responsible for more than one-quarter of the $200+ billion in annual coffee sales. The company sources coffee through the six largest exporters in the world – Brazil, Vietnam, Colombia, Indonesia, Cambodia, and Ethiopia. The sustainability team has mapped the impact of climate change across the supply chain (from growing the two predominant types of coffee plants to the end consumer drinking the coffee), but it has failed to account for the dire social conditions of the 125 million people worldwide who depend on coffee for their livelihoods, many of whom are unable to earn a reliable living. Women make up about 80% of the coffee workforce in Indonesia and 50% in Vietnam.

In both cases, these companies have failed to understand that their supply chain is compromised because their workforce is ill-equipped to anticipate, avoid or recover from climate-related disasters due to intersecting social, cultural, economic, and political inequalities. They have understood exposure to hazard but failed to appreciate the critical importance of vulnerability.

How can these companies develop the necessary skills to apply a three-dimensional approach to risk? By applying due diligence criteria to hazards, exposure, and vulnerability, as they would with other risk factors such as financial, economic, and political circumstances. In the context of climate change, this involves a four-step process to identify risks; determine potential outcome severity; clarify the scope of responsibility; and recommend actions to build resilience to the identified threats.[8]

All companies are connected to climate-vulnerable communities in some capacity, although often not in their direct operations. To determine the scope of responsibility and appropriate remedy, a company must first map its operations and the operations of significant business partners to identify where the company operates in known high-risk climate-vulnerable communities. Companies should determine whether they are causing or contributing to increasing the vulnerability of high-risk communities, or whether they are directly linked to a business partner that is increasing the vulnerability of an at-risk community. Put differently, companies should ask whether their operations (or significant business partners') cause or contribute to weakening the realization of human rights that would protect vulnerable communities from the negative impacts of climate change. For example, a food and agriculture company with a significant operational footprint in a high-risk community should take corrective measures if it learns that its business partners have acquired land without contracts, making the proper enforcement of land rights after an adverse weather event more difficult, thus weakening rights realization. Similarly, a consumer goods company should be aware of heightened risk if it is manufacturing in a region with significant levels of discrimination against women.[9]

Companies need a robust commitment to net-zero emissions

Limiting global warming to 1.5°C requires reducing emissions of CO_2 by 45% from 2010 levels by 2030 and reaching net-zero by 2050.[10] Only 10% of the estimated sixty thousand global multinational corporations are currently pursuing climate commitments; even fewer with a level of ambition in line with this target; and only 11 are deemed to have the highest and most robust type of net-zero pledge.[11] As a result, there is an urgent need for companies to revisit their commitments and ratchet up their ambition and to rapidly scale up the volume of companies making commitments. The NewClimate Institute has produced a list of basic criteria for net-zero target transparency that should guide corporate efforts:[11] These include:

- Companies should specify separate targets for emission reductions and emission removals and provide details on offsets. As illustrated in Chapter 5, too many companies are relying on excessive amounts of offsets and unproven removals technologies. This may work for a limited number of companies working in isolation but is unsustainable in terms of achieving the global goal.
- Chart a decarbonization pathway with interim targets and accountability. Too many companies have broadcast a 2050 goal without giving any clarity on what they will achieve by 2030 (interim) or the plans they will deploy over the coming five years. This makes the targets seem like rhetorical commitments rather than real action.
- Share information on emission reduction measures to facilitate good practice replication. As Chapter 4 has illustrated, investments in just six areas – solar energy, wind energy, efficient appliances, efficient cars, afforestation, and halting deforestation – would build a pathway to net-zero.[12] Project

Drawdow has identified 80 solutions that are available today that could deliver emissions reductions in these six areas.[13] Similarly, the IPCC has identified 60 so-called response options that can reduce GHG emissions from land use.[14] Although each sector and individual company will have its own tailored set of actions and innovations, this comprehensive set of solutions illustrates that the economy as a whole will be choosing from the same menu and so can learn from shared implementation lessons.

- Document stakeholder consultation approaches to demonstrate ownership of plans and specify supply chain emissions coverage to identify synergies with other companies. Collaboration and consultation will be vital to implementing climate commitments because of the importance of having a full supply-chain approach. For example, a car may contain up to 50,000 unique parts sourced from across the globe, and so developing productive relationships with all these suppliers and partners will be necessary to achieve absolute emissions reductions.[15]
- Commit to a timeline for the revision of ambition to establish an ambition ratchet mechanism. Leadership is dynamic, climate science evolves, innovation leads to new breakthroughs, and policy incentives change. Consequently, corporate commitments ought to be dynamic as well. The signatories to the Paris Agreement included a provision to review and ratchet up national climate commitments every five years with a promise to progressively increase ambition over time and avoid backsliding. Corporate commitments should align with this cycle.[16]

Make a corporate commitment to resilience

The work of building climate adaptation across global supply chains has begun. The private sector is investing in natural, human, social, physical, financial, and political capital to enhance resilience inside individual companies, across complex global supply chains, and within frontline communities. However, the work of adaptation needs to be deepened and accelerated.

According to research conducted by Yale University and the NewClimate Institute, 6,225 companies and investors from 120 countries, representing at least $36.5 trillion in revenue, have pledged at least one climate commitment.[17] These commitments are housed within a range of initiatives, including the setting of science-based emissions reductions targets; purchasing 100% of energy needs from renewable sources; ending all commodity-driven deforestation in supply chains; and reducing short-lived climate pollutants. For example, the We Mean Business Coalition invites companies to make emissions reductions pledges across 11 commitment areas including science-based targets, renewable energy procurement, and ending commodity-driven deforestation. To date, close to 1,700 companies, with market capitalization close to $25 trillion, have made over 2,100 commitments.[18] The time is ripe to complement these mitigation-focused commitments with a new pledge on climate resilience.

A resilience commitment should have a clearly stated business rationale/ purpose for focusing on resilience as well as goals among employees and supply chain partners, which can serve as guiding principles in the event of breakdowns in systems with unexpected events. Most companies have failed to analyze the full impact of climate risk on business risk factors including strategic, financial, operational, human resources, compliance, and legal risks. As a result, they do not have a full understanding of the business rationale for action. The business case can be strengthened by equipping companies with a methodology to build climate resilience backed up by documented new practices, processes, and/or investments made – highlighting opportunity as well as risk and the intended multidimensional return on investment.

Just as leadership on greenhouse gas emissions reductions is increasingly viewed through the prism of science-based targets, leadership on climate risk and resilience needs to be aligned with scientifically robust assessments of risk and resilience. This means commitments that accept the IPCC definition of climate risk (consisting of hazard, exposure, and vulnerability) rather than following an antiquated approach to risk based merely on exposure to a climate event. In addition, approaches to resilience need to be aligned with IPCC scenarios and framed around mobilizing resources in support of the six capital assets essential to adaptive capacity (natural, human, physical, social, political, and financial capital).

To succeed, businesses rely upon the availability of resources; the security of supply chains and transport routes; the availability of a reliable workforce and infrastructure; and the rising prosperity of consumers and shareholders. Consequently, an approach to risk and resilience that isolates these issues inside one company is unlikely to have much impact. Too few companies have mapped the full spectrum of risks across their full supply chain and are therefore working with flawed approaches to risk management. Commitments should, therefore, include a process of building knowledge and internal capability to understand the full implications of risk and resilience across the full value chain.

Article 4 of the Paris Agreement provides the basis for commitments that are sequenced, provided the sequencing represents a progression over time and reflects the highest possible level of ambition based on circumstances and capabilities.[16] On the mitigation side of the equation, the sequence has three steps – peaking of greenhouse gas emissions, followed by rapid reductions, and, eventually, landing at net-zero emissions in the second half of the century. A similar approach to sequencing could be applied to resilience commitments. Sequencing essentially provides an on-ramp for ambition, allowing those making the commitments to begin with a goal that is balanced between ambition and pragmatism, is achievable, and provides confidence that goals can be met but also offers a stretch goal and a roadmap for how to ultimately get there.

Conditionality is also a useful element to include in a commitment as the attainment of climate goals is often dependent on a complex ecosystem comprising governments, peers, suppliers, and consumers. Designing a goal that has elements of conditionality to it is a means for making an offer on ambition while also protecting those making the offer from being isolated and operating without a conducive enabling environment. Often, it provides a basis to balance the offer with specific asks, notably aimed at government.

What could be expected of companies that make this commitment?

1. Conduct a science-based assessment of their climate risks: The IPCC defines climate risk as determined by the existence of physical hazards, exposure to those hazards, and underlying vulnerability. With this commitment companies pledge to undertake an assessment of climate risk grounded in this definition to properly understand:
    - Projected increases in the intensity and frequency of climate hazards, including hurricanes, cyclones, changing precipitation patterns, extreme temperatures, droughts, floods, storm surges, sea-swells, salt-water intrusion, acidification of the oceans, landslides, and the spread of water-borne, vector-borne and airborne diseases as well as the spread of pests.
    - Exposure to these hazards right across the supply chain, including impacts on raw materials extraction, workforce wellbeing, manufacturing, distribution, and retail.
    - Underlying vulnerabilities inside the company, across the supply chain, and within frontline communities. This means assessing the propensity of exposed elements, whether people, ecosystems, biodiversity, economic sectors, complex supply chains, or individual companies to suffer adverse effects when exposed to climate-related physical hazards.

2. Formulate a strategy to build resilience to climate risks based on six capital assets: The IPCC defines resilience as "the ability of a system and its component parts to anticipate, absorb, accommodate, or recover from the effects of a hazardous event in a timely and efficient manner." A resilient business will, therefore, be able to anticipate, absorb, accommodate, and rapidly recover from climate events in its own operations and throughout its value chain. It will further contribute to resilient societies, which means moderating harm to socioecological systems and enabling people, the economy, and natural systems to rebound quickly in the face of adversity. With this commitment companies pledge to formulate a strategy for resilience drawing on the six capital assets – the interdependent capacities that together address the underlying causes of vulnerability. They consist of human, financial, social, natural, physical, and political capital and are considered to be the key building blocks of resilience.

3. Report climate risks and strategy to build resilience: The company will produce and use information related to climate risk and resilience in mainstream corporate reports out of a sense of fiduciary and social responsibility, in order to support the development of sound corporate strategies and the efficient allocation of capital.

The innovation agenda **139**

4. Commit to collaboration: Research conducted by Business for Social Responsibility (BSR) reveals that precompetitive collaboration allows companies to invest their resources in the sustainable development of their organization, market, and greater ecosystem. Working together to address sustainability challenges allows companies to coinvest in new market opportunities; build resilient, sustainable supply chains; overcome regulatory barriers; share the risk of new approaches with peer organizations; access donor funding to support innovation; shape industry standards; and build legitimacy and support for a preferred approach.[19]

Enable decarbonization and resilience in others

"Debank," reinvesting, and scale: shifting financial flows toward decarbonization and resilience

As Chapter 5 revealed, 35 banks from Canada, China, Europe, Japan, and the United States have together funneled $2.7 trillion into fossil fuels in the four years immediately after the adoption of the Paris Agreement (2016–2019).[20] Financial institutions must enable the transition to a low-carbon economy by taking immediate steps to debank the fossil fuel industry – establishing a process and timeline for phasing fossil fuels out of their portfolios.

A number of pioneers are providing examples of how this might be achieved. Royal Bank of Scotland (RBS) committed to a new policy in 2020 that will end financing for major oil and gas companies "unless they have credible transition plans in place by the end of 2021".[21] The $1 trillion Norwegian government pension fund, the world's largest sovereign wealth fund, has adopted a detailed set of climate-related expectations for all portfolio companies, covering strategy, risk management, disclosure, and policy. The fund has also divested its holdings in certain coal-mining and coal-burning power companies.[22] The California State Teachers' Retirement System (CalSTRS), one of the largest US public pension funds, divested from US thermal coal companies in 2016 and from non-US thermal coal companies in 2017.[23] The European Investment Bank (EIB) has gone even further, committing to phase out coal along with all unabated oil and gas projects by the end of 2021.[24] It will spend €1 trillion on a green investment package by 2030, end funding for all fossil fuels and airport expansions by the end of 2022 and target more than half of its funding activities to climate action by 2025.[25]

The US Commodity Futures Trading Commission has recommended that all relevant federal financial regulatory agencies incorporate climate-related risks into their due diligence, screening and policy posture. This would include requiring all financial institutions to embed climate risk monitoring and management into governance frameworks, and cooperating in developing climate risk stress testing for physical and transition risks.[26]

It is not enough to debank and divest; the transition to a low-carbon and resilient future will require massive reinvestment. Investment needs are broadly estimated to be in the trillions of dollars. One estimate comes from the International Renewable Energy Agency (IRENA), which charts an ambitious yet technically and economically feasible path for limiting warming to "well below" 2°C. IRENA estimates that $110 trillion of cumulative worldwide investment in the energy sector will be needed leading up to 2050. That equates to roughly 2% of average global gross domestic product (GDP) per year over the period. Of the $110 trillion, $95 trillion is already required under the reference case scenario of current plans and policies but would need to be redirected from investments in high-carbon to low-carbon activities. An additional $15 trillion is necessary to further reduce emissions.[27] This transformation is estimated to boost total global GDP by 2.5%, or 5.3% when considering the avoided climate-related damages relative to the reference case (maintenance of current plans and policies). The cumulative benefit in terms of avoided climate-related and air pollution damages ranges from $50 trillion to $142 trillion, and reducing fossil fuel subsidies would generate further savings of $15 trillion by 2050, relative to the reference case.[26]

Mobilizing this order of finance will require moving all money in the economy including public procurement, corporate procurement, and private investment. As a starting point, financial institutions can enable this by rapidly scaling up social finance. Social finance refers to any investment activity that generates financial returns and considers social and environmental impact. It encompasses four primary strategies: socially responsible investing (SRI), environmental finance, development finance, and impact investing. According to Morgan Stanley, the amount of assets under management (AUM) for sustainable investments has grown to $8.72 trillion in the United States, a 33% increase since 2014, and now constitutes 22% of the $40.3 trillion of total AUM. Millennials are driving this change – putting money in sustainable investments at a rate 2× higher than average. 86% of millennial investors say they are "very interested" or "interested" in sustainability; 61% have made at least one sustainable investment in the last year; and 75% think their investments can influence climate change. With a $30 trillion wealth transfer coming to millennials over the coming decades, this is an unprecedented opportunity to capture a generational shift in investment preferences while supporting climate resilience.[28] BNY Mellon, the largest custodian bank in the world, commissioned research that offers a series of conditions to enable mainstream investors to bring social finance to scale including:

- Increasing accessibility to attractive investment products by deepening social finance expertise across the investment value chain, particularly among advisory and investment teams.
- Improving measurement by building track records, frameworks, and presentation standards necessary to generate investor confidence including adopt standardized nonfinancial metrics across investment activities and integrating social and environmental impact into valuation and pricing of risk.

The innovation agenda 141

- Improving transparency by institutionalizing the sharing of information among investors and stakeholders as best practices to promote efficient use of capital
- Pursuing innovation and risk mitigation by leveraging expertise through partnerships.[29]

While spending on decarbonization received the majority of attention, it is also important to scale up financial support for resilience. While adaptation finance flows have increased in recent years, the current finance levels continue to fall short of both current needs and future projections. According to UNEP, current adaptation costs are likely to be at least up to three times higher than international public finance for adaptation. Adaptation costs in 2030 are likely to be in the range of $140–300 billion per year and as high as $500 billion by 2050. International public finance for adaptation in 2015 was around $9.5 billion.[30] Research commissioned by the Global Commission on Adaptation also found that investing $1.8 trillion globally in five areas from 2020 to 2030 could generate $7.1 trillion in total net benefits. The five areas the report considers are early warning systems, climate-resilient infrastructure, improved dryland agriculture, mangrove protection, and investments in making water resources more resilient.[31] The messages are clear – more finance will be needed, more will need to come from private sources, and the returns on these investments will be significant.

Choosing when to spend the money as well as how much to spend is a vital component to improving financing for climate resilience. At present, too much of the spending on climate-related disasters is focused on recovery and is therefore inflated and often far too late. Instead, early investments should be prioritized to yield the so-called "resilience dividend." The National Institute of Building Sciences estimates investing in physical assets to support climate resilience saves up to $10 per $1 invested. For example, building-code hurricane requirements save an average of $5.6 billion for an annual investment of $540 million. The Benefit-Cost-Ratios can reach as high as 30 to 1 in areas along the Gulf of Mexico and Atlantic coastline due to their exacerbated risk of more intense and frequent hurricanes.[32] Again the message is clear, early investments in resilience building offer far greater value than the cost of clean-up.

Strengthening human rights and empowering women

Chapter 1 revealed that intersecting economic, social, cultural, legal, and political inequality amplifies climate risk by exacerbating vulnerability. Vulnerable populations are often marginalized within their wider communities. They are restricted in their access to information and so cannot make informed choices about risk avoidance. They are prevented from participating in decision-making and so cannot shape public policies to enhance their resilience. They are often shut out of economic opportunity and so lack the economic resources to absorb the impact of and then recover from extreme events. Companies have long viewed human rights as the social side of the sustainability spectrum and climate change as sitting on the environmental side. Distant relatives perhaps, but with little interaction between them. This now needs to change. Climate change undermines the realization of human rights, and strengthening human rights builds climate resilience. Joining the two disciplines together therefore becomes critical to addressing both. The gender dimensions of climate change illustrate this point.

Women are disproportionately impacted by climate change as they are often constrained by social and cultural norms that prevent them from acquiring appropriate skill-sets; restrict their access to assets (including land); prevent them from having adequate access to governance (including access to decision-making and information); place them in inferior social positions; and prevent them from acquiring education and appropriate healthcare. They are often most exposed to the health risks arising from pollution, poor sanitation, and unclean water. And they often also rely mostly on natural resources, often deriving up to two-thirds of their income directly from these resources and spending up to three-quarters of their household incomes on food and other basic needs.

Across the globe, women are denied their human right to a minimum standard of living and adequate means of subsistence. Women lack access to financial services including savings, insurance, credit, and investments and are restricted in their ownership of productive assets such as property, farms, and inheritance. Women are concentrated in informal and unprotected work. Nearly 90% of African women are not covered by labor relations laws; they have no minimum wage or social protection and no maternity benefits or old age pension.[33] Women are both underpaid and unpaid: on average women earn 10–30% less than men for work of equal value.[34] The International Labour Organization (ILO) found that women devote one to four hours less a day to market activities; and, while 72% of men are employed, only 47% of women are.[35] In 2014, the Social Institutions and Gender Index (SIGI) of the Organisation for Economic Co-operation and Development (OECD) found that women had the same legal rights as men to own and access land in only 28 of the world's countries.[36] Women account for only 12.8% of agricultural landholders in the world, despite forming up to 80% of the workforce across the food and agriculture value chain.[37]

In periods of stress, women may be forced to sell off their physical assets such as household items, livestock with lower cultural importance, and jewelry. This is often after taking loans from family/friends and possibly before or after high-risk borrowing from loan sharks. Ultimately, women might be forced to sell their land, thereby undermining the sustainability of their livelihoods over the longer term.[7] A 2007 study of 141 natural disasters found that when the socioeconomic status of women is low, more women died as a result of a natural disaster; and, postdisaster, women and girls suffered a disproportionate lack of access to food and economic resources.[38] Some of the reasons for this are similar across these countries: women died because they stayed behind to look for their children and other relatives. Women in these areas often can't swim or climb trees, which meant that they couldn't escape. Recurring natural disasters also lead to further violations of women's rights and dignity, such as human trafficking, child marriage, sexual exploitation, and forced labor. The 1991 cyclone in Bangladesh illustrates many of these issues. More than 90% of the estimated 140,000 fatalities were women – their limited mobility, skills, and social status exacerbated their vulnerability to this extreme weather event.[39]

The gender dimension of climate change matters to companies because of the dependence on women workers in global supply chains. The International Labour Office calculates that approximately 190 million women work in global supply chain-related jobs in the 40 countries for which estimates were available. In sectors such as consumer products and food, the proportion of women in the labor force can be as high as 70–90% in some countries. Women farmers currently account for 45–80% of all food production in developing countries.[40] In 2017, agriculture, food, and related industries contributed more than $1 trillion, or 5.4% of GDP, and more than 3 million jobs in 2019 in the United States alone. A failure to understand the disproportionate impact of climate on women is therefore a failure to properly diagnose risk for this sector. More broadly, it is estimated that the economic costs to the global economy of discriminatory social institutions and violence against women are estimated to be approximately $12 trillion annually[41] – more than all the economic recovery and stimulus packages mobilized in response to COVID-19.

On the other hand, investing in the human rights of women leads to multiple societal and economic benefits. For example, a study using data from 219 countries from 1970 to 2009 found that, "for every one additional year of education for women of reproductive age, child mortality decreased by 9.5 percent".[42] Empowering women workers in the food and agriculture supply chain could increase yields on farms by 20–30%. This could raise total agricultural output in developing countries by 2.5–4%.[43] According to UNDP, the increased employment of women in developed economies over the past decade has contributed significantly more to global economic growth than China. Raising the female labor force participation rate to country-specific male levels is projected to raise GDP in the U.S. by as much as 5%.[44]

At minimum, corporate actions should be gender sensitive – companies should understand and take into consideration sociocultural factors underlying sex-based discrimination that amplify risk. Gender sensitive has come to mean "do no harm." Companies should work toward interventions that are gender responsive – identifying, understanding, and implementing interventions to address gender gaps and overcome historical gender biases in policies and interventions. More than "doing no harm," a gender-responsive policy, program, plan, or project aims to "do better." Ultimately, companies should aspire to climate resilience practices that are gender transformative – whereby gender is central and promoting gender equality is a priority that aims to transform unequal relations, power structures, access to and control of resources, and decision-making spheres.

The new corporate climate leader needs to show leadership in addressing structural problems that are nonclimate in nature in order to build climate resilience. For example, a failure to provide healthcare, parental leave, equal pay, secure labor contracts, land rights, and access to financial services amplify climate risk for women. The new corporate climate leaders should therefore work with governments to address these social, cultural, political, and economic inequalities and not focus exclusively on climate policy responses.

Accountability for improving the market

Shaping the policy enabling environment

In 2019, the authors of the UNDP Human Development Report used a baking metaphor to emphasize how systems failure in the market are key drivers of socioecological risks including climate change. In the introduction to the report they wrote: "the increasingly important questions for many countries are not about the overall size of the pie but the relative size of its slices. In this year's Report, though not for the first time in its history, we also worry about the oven."[45]

Even as companies commit to decarbonization, elements of the policy enabling environment may continue to pose barriers to progress. For example, in 2015, the world spent $4.7 trillion on fossil fuel subsidies, growing to $5.2 trillion in 2017 or 6.5% of GDP.[46] In addition, approximately 85% of global emissions are currently not priced, and about three-quarters of the emissions that are covered are priced well below the target price of at least $40–80/tCO$_2$ by 2020 and $50–100/tCO$_2$ by 2030, identified as being Paris-compliant by the Carbon Pricing Leadership Coalition.[47] This means that the price of fossil fuels is kept artificially low, the cost of pollution is shared by society as a whole rather than the polluter, and so there are too few price incentives to lead on climate ambition.

More than $15 trillion has already been spent on stimulus to rebuild the global economy from the COVID pandemic. In at least 18 of the world's biggest economies, rescue packages are dominated by bailouts for oil or new high-carbon infrastructure. China is the worst offender, with only 0.3% of its package – about $1.5 billion – slated for green projects. The European Union stimulus measures are the highest performers when measured against decarbonization criteria, but even only 30% of EU's $880 billion Recovery Fund is being used to advance climate-compatible policies.

Analysis of national climate action plans (so-called NDCs) by UN Climate Change shows that only 75 countries, representing 30% of GHG emissions, met the deadline for updating their plans, accounting for about 30% of global emissions. Their combined plans achieve less than 1% emissions reductions by 2030 compared to 2010 levels, at a time when the science demands 45% reductions.[48]

The new corporate climate leaders need to enhance their advocacy for these vital changes to the policy enabling environment. Equally, they need to end their lobbying for measures that undermine the transition to a low-carbon and resilient future. The very same car makers who are now working to introduce electric vehicles have been aggressively lobbying to roll back fuel efficiency standards in the United States.[49] For years, BP has promoted itself as a champion of methane, promoting an industry standard on addressing fugitive emissions reductions that, if followed by peers in the industry, would prevent 100m tonnes of methane from entering the atmosphere every year – the equivalent of cutting all the carbon dioxide emitted since the 19th century by one-sixth. To achieve the goal, BP upgraded almost 10,000 American rigs, reducing its fugitive methane emissions to 0.2% against an industry average of over 2%. And yet, at the same time, BP has lobbied intensively to weaken US rules on methane emissions and undermined its own hopes of achieving its goals as well as wider industry adoption.[50] In 2017, the Business Roundtable asked the Trump administration to make cuts to a list of top regulations considered unduly burdensome to big business. Among those were ozone and coal-fired plant admission standards. Companies taking significant climate commitments were now seeking to undermine the very policy enabling environment that would help them meet their goals.[51] This is because when corporations pursue climate commitments, they tend to stop short of measures that interfere with profits or contradict business models.[52]

The noted corporate sustainability expert John Elkington has written that there is little point in cleaning up corporate fish if we then release them back into dirty market waters.[50] Perhaps the murkiest water of all concerns the reluctance of corporate climate leaders to pay their taxes.

Acceding to corporate demands for tax and spending cuts, governments have for years been shutting down in scaling back the very social safety net that is vital to climate resilience.[51]

Nearly 100 companies in the Fortune 500 had an effective federal tax rate of 0% or less in 2018. The list of companies covers a wide range of industries and includes some of the most prominent leaders in the corporate climate movement. The report covers 379 companies from the Fortune list that were profitable in 2018 and finds that 91 paid an effective federal tax rate of 0% or less. Those companies come from a wide range of industries and include companies such as Amazon, Starbucks, Levi Strauss, and AECOM – companies with science-based emissions reductions targets. The lower average rate means that the federal government brought in about $74 billion less in corporate taxes than if all the companies had paid the statutory rate.[52]

Amazon is widely regarded as the worst offender. According to Fair Tax Mark, the company paid just $3.4 billion in tax on its income over the course of the past decade despite achieving revenues of $960.5 billion and profits of $26.8 billion. This means Amazon's effective tax rate was 12.7% over the decade when the headline tax rate in the Unied States has been 35% for most of that period.[53] The company follows the same practices overseas. In the Unied Kingdom, Amazon, without big-box stores to fund, pays less in business rates than its principal retail rivals. A UK Parliament inquiry found that while various well-known companies paid business rates at 1.5–6% of turnover, Amazon paid 0.7%. It also pays less corporation tax. In 2019 corporation tax was £14.5 million at Amazon's UK logistics company, where revenue was £2.9 billion.[54] This reputational damage to Amazon, coupled with continued questions about the integrity of its net-zero goal, comes at a moment when the company is trying to position itself as vanguard of the corporate climate community. French tax economist, Gabriel Zucman, has estimated that by 2013 about $5.9 trillion – or 10% of global GDP – was held in tax havens, with at least three-quarters of it unrecorded. Repatriating this money, even at a rate as low as 5%, would produce over $250 billion in government revenues and could offer a down-payment on climate-compatible infrastructure.[55]

Corporations are created by law and are obliged to prioritize their shareholders' interests through the best interests of the corporation rule. That rule results in the corporation having a fundamentally self-interested nature, which significantly constrains what leaders can do about climate and other sustainability goals. As Bakan has written, the law demands corporations do well, while it permits them to do good but only when that helps them do well.[51] Changing this rule, and the wider willingness of companies to sacrifice climate leadership for profit, is far more important than pushing for a new economy-wide emissions reductions target or a new energy policy. One important step will be to change the behavior of industry federations and associations, which have often been openly hostile to climate ambition. These associations are often taken hostage by high carbon interests, and because they move at the pace of their slowest and most vested members, the result is that they often actively lobby to slow or stall change.[50]

Mandatory climate disclosure to increase transparency and guide the market

The disclosure by corporations of information on climate-related financial risks is an essential building block to ensure that climate risks are measured and managed effectively. Investors can use climate-related disclosures to assess risks to firms, margins, cash flows, and valuations, allowing markets to price risk more accurately and facilitating the risk-informed allocation of capital.

In 2019, more than 630 investors managing more than $37 trillion signed the Global Investor Statement to Governments on Climate Change, which called on governments to improve climate-related financial reporting. Disclosure frameworks have been developed to enhance the quality and comparability of corporate disclosures, most notably, the Task Force on Climate-related Financial Disclosures (TCFD).

While disclosure rates are trending in a positive direction, an update published by the TCFD found significant gaps. TCFD reviewed financial reports for more than 1,700 companies in industries spanning banking, insurance, energy, media, and transportation from 69 countries. The analysis found that surveyed companies only provided, on average, 3.6 of the 11 total TCFD recommended disclosures. Only one in 15 companies disclosed information on the resilience of their strategies under different climate-related scenarios.[56] This has led US Commodities Futures Trading Commission, the existing disclosure regime has not resulted in disclosures of a scope, breadth, and quality to be sufficiently useful to investors and regulators, making it difficult for investors and others to understand exposure and manage climate risks.[26] The Commission has proposed a number of revisions to disclosure to enhance this practice as a mode of transparency and accountability:

- First, Material climate risks must be disclosed under existing law, and climate risk disclosure should cover material risks for various time horizons. To date, disclosure has been voluntary, but a number of countries including France,[57] the UK,[58] and New Zealand[59] have either made climate risk disclosure mandatory or committed to do so. Others should follow.
- Second, there should be a clear and standardized definition of materiality for disclosing medium- and long-term climate risks, including through quantitative and qualitative factors.
- Third, companies should disclose Scope 1 and 2 emissions and disclosing Scope 3 emissions should become the norm as more reliable transition risk metrics and consistent methodologies are developed.[26]

The private sector has made great strides in recent years to understand and respond to the climate crisis. However, leadership is dynamic and there is a danger that companies are resting on a 2015 model of leadership, whereby broadcasting a vague emissions reductions target was deemed sufficient for entry into the vanguard of corporate climate leaders. As a new decade begins, the time has come to embark on a new corporate climate leadership – one characterized by robust net-zero emissions reductions targets; a leadership portfolio balanced with commitments to resilience and human rights; pledges to debank the high-carbon economy while reinvesting in a low-carbon future; and a vow to reshape the market through consistent engagement with public policy and a renewed commitment to transparency.

References

1. McKibbon, B. (2017) "Winning Slowly Is the Same as Losing". *Rolling Stone Magazine*, 1 December, 2017. Los Angeles: Rolling Stone. https://www.rollingstone.com/politics/politics-news/bill-mckibben-winning-slowly-is-the-same-as-losing-198205/
2. Christensen, J. and Olhoff, A. (2019) *Lessons from a Decade of Emissions Gap Assessments*. Nairobi: United Nations Environment Programme. https://wedocs.unep.org/bitstream/handle/20.500.11822/30022/EGR10.pdf?sequence=1&isAllowed=y
3. IPCC (Intergovernmental Panel on Climate Change) (2018) *Global Warming of 1.5°C. An IPCC Special Report on the Impacts of Global Warming of 1.5°C above Pre-industrial Levels and Related Global Greenhouse Gas Emission Pathways, in the Context of Strengthening the Global Response to the Threat of Climate Change, Sustainable Development, and Efforts to*

Eradicate Poverty [Masson-Delmotte, V., P. Zhai, H.-O. Pörtner, D. Roberts, J. Skea, P.R. Shukla, A. Pirani, W. Moufouma-Okia, C. Péan, R. Pidcock, S. Connors, J.B.R. Matthews, Y. Chen, X. Zhou, M.I. Gomis, E. Lonnoy, T. Maycock, M. Tignor and T. Waterfield (eds.)]. Geneva: IPCC. https://www.ipcc.ch/site/assets/uploads/sites/2/2019/06/SR15_Full_Report_Low_Res.pdf

4. Crawford, M. and Seidel, S. (2013) *Weathering the Storm: Building Business Resilience to Climate Change*. Washington, DC: Center for Climate and Energy Solutions (C2ES). https://www.c2es.org/site/assets/uploads/2013/07/weathering-the-storm-full-report.pdf

5. Chase, M., Norton, T. and Wright, C. (2016) *"From Agreement to Action: Mobilizing Suppliers toward a Climate Resilient World"*. In: *CDP Supply Chain Report 2015/2016*. New York: BSR and CDP. https://www.bsr.org/reports/BSR_CDP_Climate_Change_Supply_Chain_Report_2015_2016.pdf

6. Goldstein, A., Turner, W., Gladstone, J. and Hole, D. (2019) "The Private Sector's Climate Change Risk and Adaptation Blind Spots". *Nature Climate Change 9*: 18–25. https://www.nature.com/articles/s41558-018-0340-5

7. Cameron, E. (2019) *Business Adaptation to Climate Change and Global Supply Chains*. The Hague: The Global Commission on Adaptation (GCA). https://static1.squarespace.com/static/59486a61e58c62bbb5b3b085/t/5d41eaa725214100017ff673/1564601001802/GCA+Supply+Chains.pdf

8. Morgan Stanley Real Assets (2018) *Weathering the Storm: Integrating Climate Resilience into Real Assets Investing*. New York: Morgan Stanley. https://www.morganstanley.com/im/publication/insights/investment-insights/ii_weatheringthestorm_us.pdf

9. Cameron, E. and Nestor, P. (2018) "Climate and Human Rights: The Business Case for Action". In: *BSR Report*. San Francisco: BSR. https://www.bsr.org/en/our-insights/report-view/climate-human-rights-the-business-case-for-action

10. IPCC. (2018) *Summary for Policymakers. In: Global Warming of 1.5°C. An IPCC Special Report on the Impacts of Global Warming of 1.5°C above Pre-Industrial Levels and Related Global Greenhouse Gas Emission Pathways, in the Context of Strengthening the Global Response to the Threat of Climate Change, Sustainable Development, and Efforts to Eradicate Poverty* [Masson-Delmotte, V., P. Zhai, H.-O. Pörtner, D. Roberts, J. Skea, P.R. Shukla, A. Pirani, W. Moufouma-Okia, C. Péan, R. Pidcock, S. Connors, J.B.R. Matthews, Y. Chen, X. Zhou, M.I. Gomis, E. Lonnoy, T. Maycock, M. Tignor, and T. Waterfield (eds.)]. In Press. Available from: https://www.ipcc.ch/sr15/download/

11. Day, T., Mooldijk, S., Kuramochi, T., Hsu, A., Yi Yeo, Z., Weinfurter, A., Tan, Y.X., French, I., Namdeo, V., Tan, O., Raghavan, S., Lim, E. and Nair, A. (2020) *Navigating the Nuances of Net-Zero Targets*. Berlin: NewClimate Institute & Data-Driven EnviroLab. https://newclimate.org/wp-content/uploads/2020/10/NewClimate_NetZeroReport_October2020.pdf

12. UNEP (United Nations Environment Programme). (2018) *The Emissions Gap Report 2018*. Nairobi: UNEP. https://www.unenvironment.org/resources/emissions-gap-report-2018

13. Hawken, P. (ed) (2017) *Drawdown: The Most Comprehensive Plan Ever Proposed to Reverse Global Warming*. New York: Penguin Books.

14. IPCC (Intergovernmental Panel on Climate Change) (2019) *Climate Change and Land: An IPCC Special Report on Climate Change, Desertification, Land Degradation, Sustainable Land Management, Food Security, and Greenhouse Gas Fluxes in Terrestrial Ecosystems* [P.R. Shukla, J. Skea, E. CalvoBuendia, V. Masson-Delmotte, H.-O. Pörtner, D. C. Roberts, P. Zhai, R. Slade, S. Connors, R. van Diemen, M. Ferrat, E. Haughey, S. Luz, S. Neogi, M. Pathak, J. Petzold, J. Portugal Pereira, P. Vyas, E. Huntley, K. Kissick, M. Belkacemi, J. Malley, (eds.)]. Geneva: IPCC. https://www.ipcc.ch/srccl-report-download-page/

15. Sheffi, Y. (2017) *The Power of Resilience: How the Best Companies Manage the Unexpected*. Cambridge, MA: The MIT Press.
16. UNFCCC (United Nations Framework Convention on Climate Change). (2015) *The Paris Agreement on Climate Change. FCCC/CP/2015/L.9/Rev.1*. Paris: UNFCCC. https://unfccc.int/sites/default/files/english_paris_agreement.pdf
17. Hsu, A., Widerberg, O., Weinfurter, A., Chan, S., Roelfsema, M., Lütkehermöller, K. and Bakhtiari, F. (2018) "Bridging the Emissions Gap – The Role of Non-state and Subnational Actors". In: *The Emissions Gap Report 2018. A UN Environment Synthesis Report*. Nairobi: United Nations Environment Programme. https://wedocs.unep.org/bitstream/handle/20.500.11822/26093/NonState_Emissions_Gap.pdf?sequence=1
18. Numbers accessed from We Mean Business Website on 11 April, 2021. www.wemeanbusinesscoalition.org
19. Enright, S., Oger, C., Pruzan-Jorgensen, P.M. and Farrag-Thibault, A. (2018) *Private-Sector Collaboration for Sustainable Development*. BSR: San Francisco. https://www.bsr.org/reports/BSR_Rockefeller_Private-Sector_Collaboration_for_Sustainable_Development.pdf
20. RAN. (2020) Banking on climate change: Fossil fuel finance report 2020. Rainforest Action Network. http://priceofoil.org/content/uploads/2020/03/Banking_on_Climate_Change_2020.pdf
21. Holger, D. (2020) "RBS Pledges End to Coal Funding, Stricter Oil Rules". In: *The Wall Street Journal (WSJ)*, 14 February 2020. New York: The Wall Street Journal. Available from: https://www.wsj.com/articles/rbs-pledges-end-to-coal-funding-stricter-oil-rules-11581716238
22. Davies, R. (2019) "Norway's $1tn Wealth Fund to Divest from Oil and Gas Exploration". In: *The Guardian*, 8 March 2019. Manchester: The Guardian. https://www.theguardian.com/world/2019/mar/08/norways-1tn-wealth-fund-to-divest-from-oil-and-gas-exploration
23. California State Teachers' Retirement System (CalSTRS). (2016) *CalSTRS Divests from U.S. Thermal Coal Companies*. West Sacramento, CA: California State Teachers' Retirement System. https://www.calstrs.com/news-release/calstrs-divests-us-thermal-coal-companies
24. Fleming, S. and Hook, L. (2019) "EIB to Phase Out Lending to Fossil Fuel Projects by 2021". *Financial Times*. https://www.ft.com/content/cc78d838-0720-11ea-a984-fbbacad9e7dd
25. European Investment bank (EIB). (2020) EU Member States Approve EIB Group Climate Bank Roadmap 2021–2025. https://www.eib.org/en/press/all/2020-307-eu-member-states-approve-eib-group-climate-bank-roadmap-2021-2025
26. Climate-Related Market Risk Subcommittee. (2020) *Managing Climate Risk in the U.S. Financial System*. Washington, DC: Commodity Futures Trading Commission, Market Risk Advisory Committee. https://www.cftc.gov/sites/default/files/2020-09/9-9-20%20Report%20of%20the%20Subcommittee%20on%20Climate-Related%20Market%20Risk%20-%20Managing%20Climate%20Risk%20in%20the%20U.S.%20-%20Financial%20System%20for%20posting.pdf
27. International Renewable Energy Agency (IRENA). (2019) *Global Energy Transformation: A Roadmap to 2050*. Abu Dhabi, UAE: International Renewable Energy Agency. https://www.irena.org/publications/2019/Apr/Global-energy-transformation-A-roadmap-to-2050-2019Edition
28. Morgan Stanley Institute for Sustainable Investing. (2017) *Sustainable Signals: New Data from the Individual Investor*. New York: Morgan Stanley & Co. LLC. https://www.morganstanley.com/pub/content/dam/msdotcom/ideas/sustainable-signals/pdf/Sustainable_Signals_Whitepaper.pdf

29. Davis Pluess, J., Govan, S. and Pelaez, P. (2015) *Conditions for Scaling Investment in Social Finance*. New York: BSR. https://www.bsr.org/reports/BSR_Conditions_for_Scaling_Social_Finance_2015.pdf
30. Olhoff A., Dickson, B., Puig, D., Alverson, K. and Bee, S. (2016) *The Adaptation Finance Gap Report*. Nairobi: United Nations Environment Programme. https://unepdtu.org/publications/the-adaptation-finance-gap-report/
31. Global Commission on Adaptation (GCA). (2019) *Adapt Now: A Global Call for Leadership on Climate Resilience*. The Hague: GCA. https://reliefweb.int/sites/reliefweb.int/files/resources/GlobalCommission_Report_FINAL.pdf
32. Porter, K. et al. (2019) *Natural Hazard Mitigation Saves: 2019 Report*. Washington, DC: Multi-Hazard Mitigation Council at the National Institute of Building Sciences. https://www.nibs.org/files/pdfs/NIBS_MMC_MitigationSaves_2019.pdf
33. Aguilar, L., Granat, M. and Owren, C. (2015) *Roots for the Future: The Landscape and Way Forward on Gender and Climate Change*. Washington, DC: IUCN & GGCA. https://wedo.org/wp-content/uploads/2015/12/Roots-for-the-future-final-1.pdf
34. ActionAid. (2015) *Close the Gap! The Cost of Inequality in Women's Work*. Johannesburg: ActionAid. https://www.actionaid.org.uk/sites/default/files/publications/womens_rights_on-line_version_2.1.pdf
35. International Labor Organization (ILO). (2014) *Global Employment Trends 2014: Risk of a Jobless Recovery?* Geneva: ILO. https://www.ilo.org/global/research/global-reports/global-employment-trends/2014/lang--en/index.htm
36. OECD. (2014). *Social Institutions and Gender Index (SIGI)*. Paris: Organization for Economic Cooperation and Development.
37. Trócaire. (2020) *Women Taking the Lead: Defending Human Rights and the Environment*. Dublin: Trócaire. https://www.trocaire.org/sites/default/files/resources/policy/women_taking_the_lead_lowres_1_0.pdf
38. Neumayer, E. and Plümper, T. (2007) "The Gendered Nature of Natural Disasters: The Impact of Catastrophic Events on the Gender Gap in Life Expectancy, 1981–2002". *Annals of the Association of American Geographers* 97(*3*): 551–566. http://eprints.lse.ac.uk/3040/1/Gendered_nature_of_natural_disasters_%28LSERO%29.pdf
39. The World Bank Group. (2011) *Gender and Climate Change: Three Things You Should Know*. Washington, DC: The World Bank Group. http://documents.worldbank.org/curated/en/274081468183862921/pdf/658420REPLACEM00Box374367B00PUBLIC0.pdf
40. ILO (International Labour Office) (2015) *World Employment and Social Outlook 2015*. Geneva: ILO. https://www.ilo.org/global/research/global-reports/weso/2015-changing-nature-of-jobs/WCMS_368626/lang--en/index.htm
41. Trócaire (2020) *Women Taking the Lead: Defending Human Rights and the Environment*. Dublin: Trócaire. Available from: https://www.ncronline.org/sites/default/files/file_attachments/Trocaire%20report%202020-Women%20Taking%20the%20Lead.pdf
42. Gakidou, E., Cowling, K., Lozano, R. and Murray, C. (2010) "Increased Educational Attainment and Its Effect on Child Mortality in 175 Countries between 1970 and 2009: A Systematic Analysis". *The Lancet* 376(*9745*): 959–974. https://pubmed.ncbi.nlm.nih.gov/22579043/
43. The Economist. (2016) "A Guide to Womenomics: Women and the World Economy". In: *The Economist*, 12 April 2016. The Economist https://www.economist.com/finance-and-economics/2006/04/12/a-guide-to-womenomics
44. UNDP (United Nations Development Programme). (2014) *Human Development Report 2014: Sustaining Human Progress: Reducing Vulnerabilities and Building Resilience*. New York: UNDP. http://hdr.undp.org/en/content/human-development-report-2014

45. United Nations Development Programme. (2020) "The Next Frontier – Human Development and the Anthropocene". In: *Human Development Report 2020*. New York: UNDP. http://hdr.undp.org/sites/default/files/hdr2020.pdf
46. Coady, D., Parry, I., Sears, L. and Shang, B. (2017) "How Large Are Global Fossil Fuel Subsidies?" *World Development* 91: 11–27. Elsevier. https://www.scopus.com/record/display.uri?eid=2-s2.0-85006306473&origin=inward&txGid=928dda79f-405fea060d1626c4587cfb1
47. High-Level Commission on Carbon Prices. (2017) *Report of the High-Level Commission on Carbon Prices*. Washington, DC: World Bank. https://www.carbonpricingleadership.org/report-of-the-highlevel-commission-on-carbon-prices
48. Farand, C. (2021) "China, US Urged to Step Up as UN Warns World 'Very Far' from Meeting Climate Goals". *Climate Home News*, 26 February, 2021. Kent: Climate Home News Ltd. https://www.climatechangenews.com/2021/02/26/china-us-urged-step-un-warns-world-far-meeting-climate-goals/
49. Meyer, R. (2018) "How the Carmakers Trumped Themselves". *The Atlantic*, 20 June, 2018. New York: The Atlantic. https://www.theatlantic.com/science/archive/2018/06/how-the-carmakers-trumped-themselves/562400/
50. Elkington, J. (2020) *Green Swans: The Coming Boom in Regenerative Capitalism*. New York: Fast Company Press.
51. Bakan, J. (2020) *The New Corporation: How "Good" Corporations Are Bad for Democracy*. New York: Vintage Books.
52. Gardner, M., Roque, L. and Wamhoff, A. (2019) *Corporate Tax Avoidance in the First Year of the Trump Tax Law*. Washington, DC: Institute on Taxation and Economic Policy (ITIP). https://itep.org/corporate-tax-avoidance-in-the-first-year-of-the-trump-tax-law/
53. Fair Tax Mark. (2019) *The Silicon Six and Their $100 Billion Tax Gap*. Manchester: Fair Tax Mark. https://fairtaxmark.net/wp-content/uploads/2019/12/Silicon-Six-Report-5-12-19.pdf
54. Editorial Board. (2020) "The Guardian View on Amazon's Dominance: We Have to Make Different Choices". In: *The Guardian*, 16 October 2020. Manchester: The Guardian. https://www.theguardian.com/commentisfree/2020/oct/16/the-guardian-view-on-amazons-dominance-we-have-to-make-different-choices
55. Zucman, G. (2013) "The Missing Wealth of Nations: Are Europe and the US Net-Debtors or Net Creditors?" *Quarterly Journal of Economics* 128(3):1321–1364. https://gabriel-zucman.eu/files/Zucman2013QJE.pdf
56. Task Force on Climate-related Financial Disclosures. (2020) *TCFD 2020 Status Report*. London: TCFD. https://www.fsb.org/2020/10/2020-status-report-task-force-on-climate-related-financial-disclosures/
57. Mazzacurati, E. (2017) "*Art. 173: France's Groundbreaking Climate Risk Reporting Law*". *Four Twenty Seven*. https://427mt.com/2017/01/16/impact-french-law-article-173/
58. Elliot, L. (2020) "UK to Make Climate Risk Reports Mandatory for Large Companies". *The Guardian*. https://www.theguardian.com/environment/2020/nov/09/uk-to-make-climate-risk-reports-mandatory-for-large-companies
59. Ministry of the Environment New Zealand (MfE). (2020) *Mandatory Climate-Related Financial Disclosures*. https://www.mfe.govt.nz/climate-change/climate-change-and-government/mandatory-climate-related-financial-disclosures

7
FINANCIAL REWARDS
Rethinking assets and liabilities in the context of climate change

Financial rewards: rethinking assets and liabilities in the context of climate change

Climate change is forcing companies to consider the financial implications of climate risk and the benefits of climate action to the bottom line. In the course of the past decade this has become increasingly common and, more importantly, increasingly out in the open. In 2018, the Nobel Prize for Economics rewarded the work of William Nordhaus, who spent his career developing models that describe and predict the relationship between economic growth and carbon emissions (Nordhaus, 2018, p. 4). In addition, during this period, we have seen the following:

- Economic models that track GDP loss to increases in temperature flourishing. A few examples include:
    - a 2012 study led by Melissa Dell at MIT estimates that a 1°C rise in temperature in a given year reduced economic growth in that year by about 1.3 percentage points in poor regions of the world (Dell, Jones and Olken 2012, p. 67);
    - a 2015 paper by Frances C. Moore and Delavane B. Diaz at Stanford tells us that temperature rise substantially slows GDP growth in poor regions (Moore and Diaz, 2015, pp. 127–131);
    - a 2015 study by Marshall Burke, Solomon M. Hsiang, and Edward Miguel at Stanford and Berkeley demonstrates that warming is expected to reshape the global economy by reducing average global incomes by roughly 23% by 2100 and widening global income inequality (Burke, Hsiang and Miguel, 2015, pp. 235–239);
    - a 2018 working paper by economists at the Federal Reserve Bank of Richmond that focused on the American economy and showed that

a one-degree F increase in average summer temperatures correlated with a decreased in the annual state-level growth rate of 0.15% to 0.25% (Colacito, Hoffman and Phan, 2019, p. 1).

- Empirical studies quantifying the economic impact of extreme weather events, with results squarely within the hundreds of billions of dollars per year (a figure which increases annually):
  - The National Oceanic and Atmospheric Administration in the United States estimated that the United States has sustained 265 weather and climate disasters since 1980 where overall damages/costs reached or exceeded $1 billion (including Consumer Price Index adjustment to 2020) and that the total cost of these 265 events exceeds $1.775 trillion (NOAA, 2021);
  - Congress in the United States appropriated $136 billion of additional funding for recovery from extreme weather events in 2017 (Painter, 2018);
  - Morgan Stanley has estimated that in 2017, property and infrastructure damage from natural disasters accounted for an estimated $220 billion in the United States, or two-thirds, of $330 billion in global economic losses (Morgan Stanley, 2018, p. 3).
- Asset managers warning us about the financial impacts of climate change:
  - Blackrock CEO Larry Fink, who sits atop almost $7 trillion in AUM, wrote in his 2020 annual letter to CEOs that climate change was going to be central to the company's investment strategy going forward, and telling his audience to be prepared for a "fundamental reshaping of finance" (Fink, 2019) indeed Blackrock estimated that under a business as usual scenario of no climate action and continued reliance on fossil fuels the risk of climate-related disasters will increase by 275% by 2050 (Schulten et al., 2019, p. 13);
  - A group of investors established in 2017, "Climate Action 100+," which, together with more than 450 investors with over $40 trillion in assets under management collectively, proposes climate progressive resolutions, and in June 2020 pushed Chevron to align with the Paris Agreement with a 53% majority (Koning Beals, 2020). This is the first time this was achieved, marking a potential turning point in the role of investors but also in the future of the fossil fuel industry.
- Insurance companies taking stock of the scale of the damage due to climate change – and of the potential opportunity – for their sector:
  - Regulators are starting to worry about this. The Bank of England's Prudential Regulation Authority, for instance, is stress-testing insurers against the risk that the world misses its carbon reduction targets (Bank of England, 2019).

154 The new corporate climate leadership

- Swiss Re estimates that insurance only covered $60 billion losses of the US$146 billion in the United States in 2019 and suggests that leaving this gap unaddressed is both a risk and an opportunity for the sector (Bevere, Gloor and Sobel, 2020, p. 1).

This type of evidence raises an important question for the finance sector: must it respond to a sea change that is already underway – the "fundamental reshaping of finance" – or must it lead this transformation? The answer, as can be expected, lies on a spectrum somewhere in between these two extremes – a spectrum that itself is also related to using GDP as the yardstick for economic and financial performance. The question, then, of whether the transformation of the finance sector is already underway will also track whether the transformation of the economy as a whole is underway. And here, too, the answer will be on a spectrum. The hypothesis driving this chapter is that those companies who are leading the "fundamental reshaping of finance" are also leading on developing new ways and innovative metrics to understand the impacts of corporate performance. In fact, it is arguable that the transformation of the economy cannot be achieved without the leadership of the financial services sector.

Specifically, this chapter will focus on understanding the push and pull of climate change on finance and how climate change impacts decision-making in the sector and reshaping their approach to assets and liabilities and to profits and loss. The first section of this chapter will present the risks of climate change to the sector, and the second section will explore the opportunities and rewards associated with ambitious climate action and leadership, arguing that in fact the framework of rewards and opportunities may be too narrow to capture the role of the financial services sector in the shift toward the new climate economy.

The value chain of the financial sector we will be working with is simplified for the purposes of this chapter (see Figure 7.1). Its purposes here is to map out

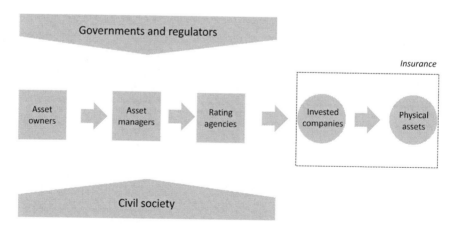

FIGURE 7.1 A sketch of the finance sector landscape

the relationships between different actors as well as the pressure points of the changing landscape of climate action and policy.

Climate-related risks to finance

Climate change poses risks to the stability of the finance sector in a number of ways that can be broken down into two major categories: (1) physical risks and (2) transition risks (TCFD, 2017).

1. Physical risks from climate change arise from a number of factors. They are the result either of specific weather events, such as heatwaves, floods, wildfires, or storms, or of longer-term changes, such as fluctuations in precipitation, extreme weather variability, sea level rise, and rising temperatures. For companies, this translates to the risk of increased business disruption and losses, as well as potential impacts on the availability and cost of capital and property and of casualty insurance. Indeed, a recent study showed that companies in developing countries most vulnerable to climate change were charged interest rates on average 0.83 percentage points higher, roughly a 10% premium (Buhr, 2018, p. 11). Furthermore, this may lead to the value of investors' portfolios fluctuating substantially and to customers paying higher insurance premiums or choosing not to buy coverage altogether, exposing them further to future risk. In addition, companies can experience physical damage to their assets such as residential and commercial property, which are often held as collateral by asset owners and asset managers. This means, for the latter, increased credit risks, and for insurers, to underwriting risks, especially if there are greater than anticipated insurance or legal claims to recover financial losses. Indeed, in the United States, the Union for Concerned Scientists estimated that, on a conservative scenario, more than 300,000 coastal homes, with a collective market value of about $117.5 billion in 2018, are at risk of chronic inundation in 2045, a timeframe that may fall within the lifespan of a 30-year mortgage, and that about 14,000 coastal commercial properties, currently assessed at a value of roughly $18.5 billion, are also at risk during that timeframe. Their total estimate is that by the end of the century, homes and commercial properties currently worth more than $1 trillion could be at risk, including the rough equivalent of all the homes in Los Angeles and Houston combined (UCS, 2018, p. 2).
2. Transition risks arise from the process of transformation of the economy toward decarbonization. Factors related to this transformation include climate-related policy and regulation, the emergence of new technology or business models, shifting views and behaviors within civil society, or evolving legal interpretations, frameworks, or precedents. For instance, transition risk occurs when minimum energy efficiency standards are implemented since they have an impact on mortgage and commercial real estate lending portfolios. Rapid technological changes lead to changes in pricing and

to a disruption in the global energy mix, as can be seen with the rise of electric vehicles (Carbon Tracker Initiative, 2018). So-called stranded assets expose entire industries to devaluation (see box below). Finally, with over 1,200 climate policies and laws in the world – representing a more than twentyfold increase since 1997 – companies in the entire economy will be exposed to policy or legal risk if they do not mitigate, adapt, or disclose their climate-related financial risks, potentially affecting their market value (Evans, 2017). This will affect not only asset owners and managers but also the insurers that have underwritten them and who provide liability cover for those companies.

STRANDED ASSETS

The potential financial consequences of climate risk are often framed in terms of "stranded assets." These are assets for which investment has already been made but will no longer be able to provide a return on that investment before the end of their economic life. That is, they will lose their value under scenarios where we tackle the climate emergency, eliciting the dilemma "either we exploit all resources and overshoot the 1.5 degree threshold, or we accept that some assets will not be exploited in order to remain below the 1.5 degree threshold." These assets are understood to be fossil fuel assets, but as we shall see below we can broaden this category to include other types of assets too. Reasons that may lead to assets being stranded include government regulation, a shift in demand, or even legal action.

Evaluating global financial assets at risk of being stranded as a result of climate action depends on scenarios and hence allows for some uncertainty or range. These factors include the date and speed at which changes are implemented, what technology is available, or the specific makeup of a company's portfolio. According to a 2015 study, a third of global oil reserves, half of gas reserves, and upwards of 80% of coal reserves are to become stranded in order to meet the targets of the Paris Agreement (McGlade and Ekins, 2015, pp. 187–190). In terms of financial value, the financial times estimated the value of those assets to amount to $900 billion in February 2020 (Livsey, 2020). Other studies have estimated that number to be in the trillions: one 2018 study presents results between $1 and 4 trillion (Mercure et al, 2018, pp. 588–593), while a 2019 estimate places that value at $2.2. trillion. These estimates vary because companies it is not in the economic interest of oil and gas companies who own the relevant data to disclose it. For comparison, the GDP of the world's fifth largest economy in 2018, the United Kingdom, was $4.8 trillion (The World Bank, 2019).

The fossil fuel sector has already started to contract, with unprecedented signals of change:

- Exxon Mobil dropped from the S&P 500 Top 10 list for the first time in nine decades in 2019 (Crowley and Kochkodin, 2019).
- As of June 2020, fossil fuel-based energy stocks had declined 34.6% for the year, and while some of this change is likely due to the COVID-19 pandemic, it also extends a decade-long trend of underperformance (Kimani, 2020). Notably, renewables stocks rose by approximately the same amount (Raval, 2020).
- As of June 2020, too, there are three companies with higher valuations than the entire fossil fuel sector, that is, higher than $1 trillion. Whereas the sector was valued at over $3 trillion, it is now worth below Apple, Microsoft, and Amazon, valued at $1.56 trillion, $1.52 trillion, and $1.36 trillion, respectively.
- In April 2020, Shell cut dividends to shareholders for the first time since World War II (Bousso and Nasralla, 2020). Also that month, oil and gas prices went negative for the first time in history – and while this can be attributed to the COVID-19 pandemic's impact on global oil demand (Walker, 2020), it also highlights the vulnerability of the sector that had been weakening for a decade before this event.
- As of July 1 2020, it is estimated that the growth in electric vehicles could bring about peak oil by the late 2020s (Carbon Tracker Initiative, 2018).
- In June 2020, oil major BP devalued its oil and gas assets by $17.5 billion (Sanderson, 2019) since it forecast that the economic context would accelerate the transition to renewables. That month, it also sold its petrochemicals business for $5 billion (Jolly, 2020).
- Less than half of the companies in the oil and gas sector carry the US Fed's minimum credit rating of BBB – according to data from S&P global. Of these, nearly half fall just on the line or slightly above it. Only one coal company has a rating above the US Fed's minimum. As a reminder, a company's credit rating describes how likely an investor in a given company is likely to be repaid. This suggests that these companies are at risk of bankruptcy (Passwaters, 2020).

Certainly, the effects of the coronavirus pandemic of 2020 can explain much of this change, but these signals would not have occurred had the transition away from fossil fuels been on its way before it hit. At the time of writing, we cannot predict to what extent this transition will have accelerated as a result, but we can observe that given the prominence of the fossil fuel sector in the economy, repercussions will be felt throughout as the industry sunsets or transforms in line with the decarbonization of the economy.

TABLE 7.1 Economics sectors and their exposition to climate-related finance risk

Fossil fuel industry sectors	Heavy-energy use and high emitting sectors	Carbon-exposed and climate-risk exposed sectors	Carbon sinks (that may also be high-emitting)	Emerging-to-established sectors leading the transition
• Coal mining and transportation • Oil and gas including petrochemicals • Natural resource extraction • Power generation and utilities	• Construction • Transportation • Chemicals • Steel • Industrial manufacturing • Agriculture (can also be a carbon sink) • Forestry (can also be a carbon sink)	• Financial services • Pharma and healthcare • Retail • Fast moving consumer goods • IT and telecoms • Real estate	• Agriculture • Forestry	• Manufacture of renewables assets • Energy efficiency technologies • Batteries and other forms of energy storage

Transition risks are associated with the risk of stranding assets, or of shortening the lifecycle of assets as a result of policy changes or other so-called externalities. It is not only the oil reserves of fossil fuel companies that run the risk of becoming stranded assets as the world moves toward decarbonization. Indeed, companies in all sectors may find that they own assets that will become worthless if we move toward a decarbonized economy. Every sector uses energy and most companies are exposed to carbon. Each company's exposure depends on idiosyncratic factors such as business model, strategy, locations, assets, and liabilities. Table 7.1 shows a spectrum of risk exposure from the most exposition to the least in the context of the stranding of assets and climate-related finance risk.

In addition to the direct risks due to physical and transition factors, companies are also exposed to the negative feedback from tighter financial conditions that the financial services sectors put in place as a response. Business disruption, asset destruction, migration, reconstruction, stranded assets, and increases in energy prices all lower corporate asset value, household wealth, and corporate profits, which in turn creates losses in the financial system, which in turn must recoup them by tightening conditions on the economy. The financial risks of climate change are exacerbated as they reverberate in the finance value chain. Indeed, we have already seen that some companies in the riskiest regions of the world are subject to higher mortgage premiums. The same feedback phenomenon is observed with insurance premiums as more losses are underwritten and as risk increases. In the aftermath of bush fires in Australia in 2019–2020, QBE, an Australian insurance company, said that climate change would make some premiums unaffordable, in particular for clients who are exposed to extreme weather events (Reuters, 2020). Munich Re, one of the world's largest reinsurers, made the same observation after the California fires of 2019 and on the basis of decades-long analysis of climate-related weather events, adding that the unaffordability of insurance could become a social issue too (Sheehan, 2019).

Climate-related opportunities for finance

Investments required for addressing climate change are estimated in the trillions of US dollars, with investments in infrastructure alone requiring about $6.9 trillion per year up to 2030 (OECD, 2018, p. 1). The New Climate Economy estimates that "bold [climate] action could yield a direct economic gain of $26 trillion through to 2030 compared with business-as-usual" (Mountford et al., 2018, p. 8). From this point of view, climate change represents for the financial sector as much a source of opportunity as a source of risk. We outlined in Chapter 2 some of the economy-wide opportunities stemming from the transition to decarbonization, broken down by sector. For companies in the finance sector, understood to be captured by Figure 7.1, the nature of opportunities is somewhat different. The finance sector is first and foremost a potential *enabler* of the transition, and it is as such that it can demonstrate leadership. We can certainly expect that finance sector companies that avail of these opportunities to *enable* will also reap the fruit of financial opportunity – indeed, we cited evidence in the previous section that climate change risk has justified, for some banks, charging higher loan premiums. But what we are talking about here are opportunities for leadership and not only for economic gain.

There are several types of such opportunities, which we will present below. They consist in taking a catalytic role within the economy either through financing projects, adapting lending policies, or focusing on adaptation and resilience.

While there is some positive momentum building among financial institutions, these activities need to become mainstream if we are to reach decarbonization at pace and at scale – and start to address the systemic issues hampering this transition. This kind of change relies on This partnership not only within the finance sector but also between these organizations, policy makers, and civil society. The opportunities for leadership listed below all rely on this kind of partnership.

Opportunities for leadership through "sustainable finance"

The growth of sustainable finance, which integrates environmental, social, and governance criteria into investment decisions across all asset classes under the heading "ESG" shows that investors are increasingly concerned with so-called extra-financial factors in decision-making. This includes climate change. Socially aware and responsible investing is at least half a century old, with origins in the 1970s, yet definitions and criteria are not broadly uniform. As a result, estimates of the value of the entire ESG stock on earth vary from $3 trillion (J.P. Morgan, 2021'\) to $31 trillion (Global Sustainable Investment Alliance, 2019, p. 3). One thing that is constant in all analyses is that this area of investment is growing. For instance, green bonds had grown to an estimated $590 billion in August 2019 from

$78 billion in 2015 (International Monetary Fund, 2019, p. 85). According to a 2020 Moody's survey, asset managers that consider ESG a top three priority have an average AUM replacement rate of 117% over the past three years. In comparison for those that did not it was just 73% (Moody's, 2020, p. 4). Furthermore, Moody's estimates that the growth potential for the ESG market is large, with a potential addressable market for ESG products of $89 trillion (Moody's, 2020, p. 1).

Signals of mainstreaming are starting to accrue and suggest that ESG is not only growing, but it is expected to grow fast. A 2017 survey by KPMG shows that 75% of the largest 100 companies across 49 countries say they are using ESG business models or incorporating aspects of sustainability approaches compared with 12% in 1993 (Blasco and King, 2017, p. 9). According to the 2018 UBS Investor Watch Global Survey, 82% of respondents said sustainable companies were good investments because they were forward-thinking and better managed (UBS, 2018, p. 3). The same study showed that investors across all ages, wealth levels, and regions said sustainable investing was growing in importance and that 58% expected it to become the "new normal" in a decade (UBS, 2019). ESG investments are proved to have satisfactory returns, sometimes outperforming more traditional investments. The MSCI ACWI ESG Leaders Index has surged more than 50% between 2015 and 2020, beating the advance of about 35% in the MSCI All-Country World Index (Mookerjee and Vishnoi, 2020). Finally, it has become evident that ignoring so-called non-financial risks can lead to sharp consequences: the COVID-19 pandemic of 2020 has spurred a distinct increase in ESG investing. In the first four months of 2020, investors channeled more than double the amount they did a year prior toward ESG funds (McCabe, 2020). The 2008/2009 financial crisis saw an uptick in membership to the UN Global Compact and the UN Principles of Responsible Investing, and they are seeing similar patterns in the midst of the COVID-19 pandemic (Kell, 2020). Indeed, both of these outlets allow companies to build resilience, build trust, and increase chances of a better recovery (Reeves et al., 2020).

Notwithstanding its growth potential, ESG investment remains small relative to the entire market, and it remains to be seen whether these trends – especially the ones spurred by the 2020 pandemic – will further consolidate into changing the shape of the entire market. At the time of writing, we are cautiously hopeful that genuine shifts are occurring, but we cannot predict whether they will be of the scale necessary to shift the entire economy toward decarbonization.

Opportunities for leadership through changing lending policies

Banks are starting to adjust their lending policies by, for example, giving discounts on loans for sustainable projects. There are tens of examples of such policies, and some notable initiatives and strategies include:

- European banks partnering with IFC and GEF to support renewable energy and energy efficiency projects generating investment of $330 million across

829 projects in Eastern Europe in a program called Commercializing Energy Efficiency Finance (CEEF) (Wang et al., 2013).
- Chinese banks participating in China Utility-Based Energy Efficiency Finance Program provided loans worth $790 million, financing 226 projects and reducing emissions by 19 million tons of carbon dioxide/year (IFC, 2011).
- The global banking firm ING Group setting a target in 2017 so that by the end of 2025, they will no longer finance clients in the utilities sector that are over 5% reliant on coal-fired power in their energy mix while continuing to finance non-coal energy projects for these clients in support of their energy transition. For new clients, the bank will support new clients in the utilities sector only when their reliance on coal is 10% or less and they have a strategy to reduce their coal percentage to close to zero by 2025. In 2017, 60% of the bank's utilities financing was in renewables, amounting to EUR 29 billion (ING, 2017).
- French bank BNP Paribas BNP Paribas has decided to cease all financing related to the thermal coal sector in 2030 in the European Union and worldwide by 2040. And to strengthen its support for the development of renewable energies, with a new financing target of EUR 18 billion by 2021 (BNP Paribas, 2019).
- In 2016, HSBC stopped the financing of new thermal coal mines and new customers dependent on thermal coal mining and had committed to investing $1 billion in high-quality liquid assets covering a range of low-carbon initiatives (HSBC, 2020). This is not strictly speaking the kind of policy that constitutes leadership since HSBC continues to finance coal power and does not have a phase-out plan, and since it does not state a clear policy on low-carbon initiatives, however, we take this to be a first step toward leadership that more timid banks could emulate.

To date, over 100 banks and institutional investors have strategies to exit thermal coal financing and shift their investments toward renewables. Not all divestment strategies are paired with favorable loan policies, however (IEEFA, 2019). The opportunity for leadership lies in this pairing, i.e. not only in divesting from an industry that is under stress but also turbo-charging the shift toward non-fossil-fuel-based power sources and generation as well as other low-carbon projects. This will no doubt rely on enabling policy, especially until this shift becomes mainstream within the banking industry.

Opportunities for leadership through adaptation finance and investing

The finance sector's engagement in adaptation is currently marginal. The main sources of finance come from UNFCCC umbrella groups such as the Adaptation Fund, supported by governments, development banks, and multilateral banks.

These funds are often linked to development aid, constituting a barrier to private sector involvement.

There are nonetheless examples of opportunities here too that signal a change in patterns. Insurance and reinsurance companies are the most engaged organizations in adaptations; however, the need for private sector investing and financing in infrastructure adaptation is becoming more urgent. Some examples of financial sector organizations availing of opportunities related to adaptation include:

- The German insurance company Allianz has developed a catastrophe bond and index-based crop insurance (Simon, 2019), and offers micro-insurance products against climate impacts (Hermann, Koferl and Mairhöfer, 2016, p. 3).
- HSBC released the HSBC Climate Vulnerability Assessment, which maps risk for the G20 in 2020 related to climate impacts in terms of food losses, water stress, and rising healthcare costs. This is intended as an advisory product for clients but also to help shape future products. This led to the development of an offering for Brazilian farmers in partnership with Allianz, to help them deal with climate-related losses (UNFCCC, 2011). In addition, HSBC has established a climate change research program with the UK Met Office to make more accurate assessments of the climate risks and impacts across investment portfolios. This is part of a major effort by HSBC to mainstream understanding of the broader impacts of climate change on investments (Ilett, 2008).

Climate action beyond climate

The finance sector as well as companies in other sectors contribute to decarbonization not only through direct financial flows but by bringing about the conditions that will enable financial markets to embed climate change into how they price risk and make financing and investment decisions. What is more, becoming an enabling – or better still, catalytic – force in the economy will require changes that go beyond a climate strategy and climate action. Of course, first and foremost, companies are to design climate strategies that address climate-related risk and implement them. The Bank of England, for one, had already told firms to establish a plan by October 2019 to mitigate climate-related risks, both physical and transition, and has even given them a deadline to do so by the end of 2021. As this suggests, the window for companies to do so is small, and they must act quickly. These strategies will inevitably include changes in investment and financing policies, as outlined in the previous section.

In addition, there are other steps companies can take to manage the risks outlined in this chapter and to avail of opportunities for leadership related to finance specifically. Currently, actions that allocate capital to low-carbon, energy-efficient opportunities, reduce portfolio emissions through implementing fossil fuel reduction policies, improve carbon emissions measurement and transparency, participate in the positive climate policy lobbying are all key. In addition, the following list presents actions that will not only accelerate the shift to decarbonization but favor the scale and pace necessary for it to be achieved successfully:

1. Companies in all sectors including finance can build capacity and leadership to shape robust investments plans that align with sector decarbonization policies and that facilitate the flow of private sector investment toward climate-proof infrastructure.
2. Companies in all sectors including finance can accelerate investment in climate-proof infrastructure, logistics, and innovation by integrating climate action at the heart of growth strategies and investment plans. This will most likely involve collaboration with the public sector, such as multilateral banks.
3. Companies in the finance sector in particular can support the implementation of favorable policies for the decarbonization of the economy by providing incentives, for instance. Carbon pricing is now in place or planned in 57 jurisdictions as of April 2019 (Carbon Pricing Leadership Coalition, 2019), but in most places it is priced too low to drive transformational change. Broadening the adoption of carbon pricing and increasing the price so that it is material to decision-making is key to making this policy instrument effective. Investors and banks can positively lobby for this.
4. Companies in all sectors, and investors in particular, can push for the implementation of the recommendations of the Task Force on Climate-Related Financial Disclosure (TCFD) on a broad scale. This favors the transparency that is necessary for investors to understand the risks of current investments and the opportunities of shifting toward decarbonization.
5. Companies can support supply chain transparency by favoring a shift to low-carbon supply chains and circular economy models through positive lobbying against subsidies, tax breaks, and regulations that hamper that shift. In addition, they can push for supply chain innovation via investment and financing toward decarbonization. Some successes have been seen in forestry, for instance (Neeff and Linhares-Juvenal, 2016).
6. Companies must ensure that human welfare and equity are at the center of decarbonization strategies to eschew inefficient, inequitable, and polluting models of the past. Just transition plans will favor the stability of the transition and make its long-term success more likely.

These types of initiatives will have aim at addressing the structural barriers that remain high in the area of climate action, particularly the uncertainty around regulation and policy. These barriers have consequences on technology development and deployment and the physical impacts of climate change due to delayed action, particularly in developing countries. These knock-on effects increase the cost of financing and investing in low carbon opportunities in the medium and long term, making rapid action the most effective way to address these issues. Every day that an opportunity to lead is not taken is a day that makes the task much more burdensome and much more uncertain.

References

Bank of England (2019) *General Insurance Stress Test 2019*. Available at: https://www.bankofengland.co.uk/-/media/boe/files/prudential-regulation/letter/2019/general-insurance-stress-test-2019-scenario-specification-guidelines-and-instructions.pdf (Accessed: 20 March 2021)

Bevere, L., Gloor, M. and Sobel, A. (2020) *Sigma Natural Catastrophes in Times of Economic Accumulation and Climate Change*. Available at: https://www.swissre.com/dam/jcr:85598d6e-b5b5-4d4b-971e-5fc9eee143fb/sigma%202%202020%20_EN.pdf (Accessed: 20 March 2021)

Blasco, J.L. and King, A. (2017) *The Road Ahead. The KPMG Survey. Of Corporate Responsibility Reporting 2017*. Available at: https://integratedreporting.org/wp-content/uploads/2017/10/kpmg-survey-of-corporate-responsibility-reporting-2017.pdf (Accessed: 20 March 2021)

BNP Paribas (2019) *BNP Paribas Announces a Timeframe for a Complete Coal Exit and Raises Its Financing Targets for Renewable Energies* [Press release], 22 November. Available at: https://group.bnpparibas/en/press-release/bnp-paribas-announces-timeframe-complete-coal-exit-raises-financing-targets-renewable-energies (Accessed: 20 March 2021)

Bousso, R. and Nasralla, S. (2020) 'Shell Cuts Dividend for First Time Since World War Two,' *Reuters*, 30 April [Online]. Available at: https://www.reuters.com/article/us-shell-results/shell-cuts-dividend-for-first-time-since-world-war-two-idUSKBN22C0TK (Accessed: 20 March 2021)

Buhr et al. (2018) Climate Change and the Cost of Capital in Developing Countries. Available at: http://unepinquiry.org/wp-content/uploads/2018/07/Climate_Change_and_the_Cost_of_Capital_in_Developing_Countries.pdf (Accessed: 20 March 2021)

Burke, M., Hsiang, S.M. and Miguel, E. (2015) 'Global Non-Linear Effect of Temperature on Economic Production,' *Nature*, 527 [online]. Available at: https://www.nature.com/articles/nature15725 (Accessed: 25 March 2021)

Carbon Pricing Leadership Coalition (2019) *Carbon Pricing in Action* [Online]. Carbon Pricing Leadership. Available at: https://www.carbonpricingleadership.org/who (Accessed: 20 March 2021)

Carbon Tracker Initiative (2018) 'Electric Vehicles: The Catalyst to Further Decarbonisation – Interactive Tool,' *Carbon Tracker*, 1 July [Online]. Available at: https://carbontracker.org/reports/ev-oil-demand-displacement-tool/ (Accessed: 20 March 2021)

Colacito, R., Hoffman, B. and Phan, T. (2019) *Temperature and Growth: A Panel Analysis of the United States*. Federal Reserve Bank of Richmond Working Paper No 18-09. Available at: https://www.richmondfed.org/-/media/richmondfedorg/publications/research/working_papers/2018/pdf/wp18-09.pdf (Accessed: 20 March 2021)

Crowley, K. and Kochkodin, B. (2019) 'Exxon Poised to Drop From S&P 500's Top 10 for First Time Ever,' *Bloomberg*, 30 August [Online]. Available at: https://www.bloomberg.com/news/articles/2019-08-30/exxon-poised-to-drop-from-s-p-500-s-top-10-for-first-time-ever (Accessed: 20 March 2021)

Dell, M., Jones, B.F. and Olken, B.A. (2012) 'Temperature Shocks and Economic Growth: Evidence from the Last Half Century,' *American Economic Journal: Macroeconomics*, 4(3) [online]. Available at: https://scholar.harvard.edu/files/dell/files/aej_temperature.pdf (Accessed: 25 March 2021)

Evans, E. (2017) 'Mapped: Climate Change Laws around the World,' *Carbon Brief*, 11 May [Online]. Available at: https://www.carbonbrief.org/mapped-climate-change-laws-around-world (Accessed: 20 March 2021)

Fink, L. (2019) *Larry Fink's 2019 Letter to CEOs. Purpose & Profit* [Online]. BlackRock. Available at: https://www.blackrock.com/corporate/investor-relations/2019-larry-fink-ceo-letter (Accessed: 20 March 2021)

Global Sustainable Investment Alliance (2019) *2018 Global Sustainable Investment Review.* Available at: http://www.gsi-alliance.org/wp-content/uploads/2019/03/GSIR_Review 2018.3.28.pdf (Accessed: 20 March 2021)

Hermann, A., Koferl, P. and Mairhöfer, J.P. (2016) *Climate Risk Insurance: New Approaches and Schemes. Allianz Working Paper.* Available at: https://www.allianz.com/content/dam/onemarketing/azcom/Allianz_com/migration/media/economic_research/publications/working_papers/en/ClimateRisk.pdf (Accessed: 20 March 2021)

HSBC (2020) *ESG Reporting and Policies* [Online]. HSBC. Available at: https://www.hsbc.com/who-we-are/esg-and-responsible-business/esg-reporting-and-policies (Accessed: 20 March 2021)

IEEFA (2019) *Financial Institutions Are Restricting Thermal Coal Funding* [Online]. IEEFA. Available at: https://ieefa.org/finance-exiting-coal/ (Accessed: 20 March 2021)

IFC (2011) *China Utility-based Energy Efficiency Finance Program (CHUEE)* [Online]. Available at: https://www.ifc.org/wps/wcm/connect/43fc31b8-9085-48c1-b92d-496ae60b19a3/Chuee+brochure-English-A4.pdf?MOD=AJPERES&CVID=jZVzScX (Accessed: 20 March 2021)

Ilett, D. (2008) 'HSBC and Met Office Team Up on Climate Change Research,' *Greenbang* [Online]. Available at: https://greenbang.com/hsbc-and-met-office-team-up-on-climate-change-research_5731.html# (Accessed: 20 March 2021)

ING (2017) 'ING Further Sharpens Coal Policy to Support Transition to Low-Carbon Economy,' *ING*, 12 December [Online]. Available at: https://www.ing.com/Newsroom/News/ING-further-sharpens-coal-policy-to-support-transition-to-low-carbon-economy.htm (Accessed: 20 March 2021)

International Monetary Fund (2019) *Global Financial Stability Report: Lower for Longer.* Available at: https://www.imf.org/en/Publications/GFSR/Issues/2019/10/01/global-financial-stability-report-october-2019 (Accessed: 20 March 2021)

Jolly, J. (2020) 'BP Sells Petrochemical Business to Ineos for $5bn,' *The Guardian*, 29 June [Online]. Available at: https://www.theguardian.com/business/2020/jun/29/bp-sells-petrochemical-business-to-ineos-for-5bn-jim-ratcliffe-plastics (Accessed: 20 March 2021)

Kell, G. (2020) 'Covid-19 Is Accelerating ESG Investing and Corporate Sustainability Practices,' *Forbes*, 19 May [Online]. Available at: https://www.forbes.com/sites/georgkell/2020/05/19/covid-19-is-accelerating-esg-investing-and-corporate-sustainability-practices/?sh=e3af5d026bbc (Accessed: 20 March 2021)

Kimani, A. (2020) 'Three Companies That Are Bigger Than the Entire Oil & Gas Industry,' *Oil Price*, 23 June [Online]. Available at: https://oilprice.com/Energy/Energy-General/Three-Companies-That-Are-Bigger-Than-The-Entire-Oil-Gas-Industry.html (Accessed: 20 March 2021)

Koning Beals, R. (2020) 'For First Time Ever, Majority of Shareholders Push Oil Giant Chevron to Align with Paris Climate Pact,' *MarketWatch*, 24 June [Online]. Available at: https://www.marketwatch.com/story/for-first-time-ever-majority-of-shareholders-push-oil-giant-chevron-to-align-with-paris-climate-pact-2020-06-23 (Accessed: 20 March 2021)

Livsey, A. (2020) 'Lex in Depth: The $900bn Cost of "Stranded Energy Asset,"' *Financial Times*, 4 February [Online]. Available at: https://www.ft.com/content/95efca74-4299-11ea-a43a-c4b328d9061c (Accessed: 20 March 2021)

McCabe, C. (2020) 'ESG Investing Shines in Market Turmoil, with Help from Big Tech,' *The Wall Street Journal,* 12 May [Online]. Available at: https://www.wsj.com/articles/esg-investing-shines-in-market-turmoil-with-help-from-big-tech-11589275801 (Accessed: 20 March 2021)

McGlade, C. and Ekins, P. (2015) 'The Geographical Distribution of Fossil Fuels Unused When Limiting Global Warming to 2°C', *Nature,* 517 [Online]. Available at: https://www.nature.com/articles/nature14016#citeas (Accessed: 20 March 2021)

Mercure, J.-F. et al. (2018) 'Macroeconomic Impact of Stranded Fossil Fuel Assets,' *Nature,* 8 [Online]. Available at: https://www.nature.com/articles/s41558-018-0182-1 (Accessed: 20 March 2021)

Moody's (2020) *Beyond Passive, ESG Investing Is the Next Growth Frontier for Asset Managers.* Available at: https://www.eticanews.it/wp-content/uploads/2020/03/Moodys_Sector-In-Depth-Asset-Managers-Global-27Feb20.pdf (Accessed: 20 March 2021)

Mookerjee, I. and Vishnoi, A. (2020) 'MSCI Says ESG Indexes Will Be Bigger Than Traditional Gauges,' *Bloomberg,* 13 February [Online]. Available at: https://www.bloomberg.com/news/articles/2020-02-13/msci-says-esg-indexes-will-be-bigger-than-traditional-benchmarks (Accessed: 20 March 2021)

Moore, F.C. and Diaz, B.D. (2015) 'Temperature Impacts on Economic Growth Warrant Stringent Mitigation Policy,' *Nature Climate Change,* 5 [Online]. Available at: https://www.nature.com/articles/nclimate2481 (Accessed: 25 March 2021)

Morgan, J.P. (2021) *JESG Investing 2021: Momentum Moves Mainstream.* New York. Available at: https://www.jpmorgan.com/insights/research/build-back-better-esg-investing (Accessed: 1 September 2021)

Morgan, Stanley (2018) *Weathering the Storm: Integrating Climate Resilience into Real Assets Investing.* Available at: https://www.morganstanley.com/im/publication/insights/investment-insights/ii_weatheringthestorm_us.pdf (Accessed: 20 March 2021)

Mountford, H. et al. (2018) *Unlocking the Inclusive Growth Story of the 21st Century: Accelerating Climate Action in Urgent.* Available at: https://newclimateeconomy.report/2018/wp-content/uploads/sites/6/2019/04/NCE_2018Report_Full_FINAL.pdf (Accessed: 20 March 2021)

Neeff, T. and Linhares-Juvenal, T. (2016) *Zero Deforestation Initiatives and Their Impacts on Commodity Supply Chains.* Available at: http://www.fao.org/3/i6857e/i6857e.pdf (Accessed: 20 March 2021)

NOAA National Centers for Environmental Information (NCEI) (2021) Billion-Dollar Weather and Climate Disasters: Overview [Online]. NOAA. Available at: https://www.ncdc.noaa.gov/billions/ (Accessed: 20 March 2021)

Nordhaus, W.D. (2018) 'Climate Change: The Ultimate Challenge for Economics' [PowerPoint presentation]. *Nobel Lecture in Economic Sciences.* Available at: https://www.nobelprize.org/uploads/2018/10/nordhaus-slides.pdf (Accessed: 20 March 2021)

OECD (2018) *Financing Climate Futures Rethinking Infrastructure.* Available at: https://www.oecd.org/environment/cc/climate-futures/policy-highlights-financing-climate-futures.pdf (Accessed: 20 March 2021)

Painter, W.L. (2018) *2017 Disaster Supplemental Appropriations: Overview.* Available at: https://fas.org/sgp/crs/homesec/R45084.pdf (Accessed: 20 March 2021)

Passwaters, M. (2020) 'Data, Analysts Point to Long Line of Struggling Firms After Chesapeake Energy,' *S&P Global,* 2 July [Online]. Available at: https://www.spglobal.com/marketintelligence/en/news-insights/latest-news-headlines/data-analysts-point-to-long-line-of-struggling-firms-after-chesapeake-energy-59265153 (Accessed: 20 March 2021)

Raval, A. (2020) 'BP to Take up to $17.5bn Hit on Assets After Cutting Energy Price Outlook,' *Financial Times*, 15 June [Online]. Available at: https://www.ft.com/content/2d84fc23-f38d-498f-9065-598f47e1ea09 (Accessed: 20 March 2021)

Reeves, M. et al. (2020) 'Emerging Strategy Lessons from COVID-19,' *BCG Henderson Institute*, 12 April [Online]. Available at: https://bcghendersoninstitute.com/emerging-strategy-lessons-from-covid-19-c1e5f9a7ba83 (Accessed: 20 March 2021)

Reuters (2020) 'Climate Change Could Make Premiums Unaffordable: QBE Insurance,' *Reuters*, 17 February [Online]. Available at: https://www.reuters.com/article/us-climate-change-qbe-ins-grp/climate-change-could-make-premiums-unaffordable-qbe-insurance-idUSKBN20B0DA (Accessed: 20 March 2021)

Sanderson, H. (2019) 'Clean Energy Shares Streak Ahead of Fossil Fuel Stocks,' *Financial Times*, 2 October [Online]. Available at: https://www.ft.com/content/2586fa10-e122-11e9-b112-9624ec9edc59 (Accessed: 20 March 2021)

Schulten, A. et al. (2019) *Getting Physical Scenario Analysis for Assessing Climate-Related Risks*. Available at: https://www.blackrock.com/us/individual/literature/whitepaper/bii-physical-climate-risks-april-2019.pdf (Accessed: 20 March 2021)

Sheehan, M. (2019) 'Wildfire Premiums Could Become a "Social Issue": Munich Re Climatologist,' *Reinsurance News*, 8 April [Online]. Available at: https://www.reinsurancene.ws/wildfire-premiums-could-become-a-social-issue-munich-re-climatologist/ (Accessed: 20 March 2021)

Simon, S. (2019) '3-Year Partnership Signed with Allianz Africa,' *Oko*, 5 August [Online]. Available at: https://www.oko.finance/post/2019/08/05/3-year-partnership-signed-with-allianz-africa (Accessed: 20 March 2021)

TCFD (2017) *Final Report. Recommendations of the Task Force on Climate-Related Financial Disclosures*. Available at: https://assets.bbhub.io/company/sites/60/2020/10/FINAL-2017-TCFD-Report-11052018.pdf (Accessed: 20 March 2021)

The World Bank (2019) GDP (Current US$) [Online]. Data The World Bank. Available at: https://data.worldbank.org/indicator/NY.GDP.MKTP.CD (Accessed: 20 March 2021)

UBS (2018) *UBS Investor Watch. Return on Values*. Available at: https://www.ubs.com/content/dam/ubs/microsites/ubs-investor-watch/IW-09-2018/return-on-value-global-report-final.pdf (Accessed: 20 March 2021)

UBS (2019) 'Decisions, Decisions on the Year (and Decade) Ahead,' *UBS*, 12 November [Online]. Available at: https://www.ubs.com/microsites/wma/insights/en/investor-watch/2019/decisions.html (Accessed: 20 March 2021)

UCS – Union of Concerned Scientists (2018) *Underwater Rising Seas, Chronic Floods, and the Implications for US Coastal Real Estate*. Available at: https://www.ucsusa.org/sites/default/files/attach/2018/06/underwater-analysis-full-report.pdf (Accessed: 20 March 2021)

UNFCCC (2011) *New Insurance Products and Climate RISK* [Online]. UNFCCC. Available at: https://www4.unfccc.int/sites/NWPStaging/Pages/item.aspx?ListItemId=24076&ListUrl=/sites/NWPStaging/Lists/MainDB (Accessed: 20 March 2021)

Walker, A. (2020) 'US Oil Prices Turn Negative as Demand Dries Up,' *BBC*, 21 April [Online]. Available at: https://www.bbc.com/news/business-52350082 (Accessed: 20 March 2021)

Wang, X. et al. (2013) 'Case Study: Commercializing Energy Efficiency Finance (CEEF)' in Wang, X. et al. (eds.) *Unlocking Commercial Financing for Clean Energy in East Asia* [Online]. Available at: https://elibrary.worldbank.org/doi/pdf/10.1596/9781464800207_Ch18 (Accessed: 20 March 2021)

8
CORPORATE GOVERNANCE, PURPOSE, AND STAKEHOLDERS

Transforming companies to transform the economy

While each of our individual companies serves its own corporate purpose, we share a fundamental commitment to all of our stakeholders. We commit to:

- Delivering value to our customers. We will further the tradition of American companies leading the way in meeting or exceeding customer expectations.
- Investing in our employees. This starts with compensating them fairly and providing important benefits. It also includes supporting them through training and education that help develop new skills for a rapidly changing world. We foster diversity and inclusion, dignity and respect.
- Dealing fairly and ethically with our suppliers. We are dedicated to serving as good partners to the other companies, large and small, that help us meet our missions.
- Supporting the communities in which we work. We respect the people in our communities and protect the environment by embracing sustainable practices across our businesses.
- Generating long-term value for shareholders, who provide the capital that allows companies to invest, grow and innovate. We are committed to transparency and effective engagement with shareholders.

Each of our stakeholders is essential. We commit to deliver value to all of them, for the future success of our companies, our communities and our country.

> Business Roundtable, "Statement of the Purpose of a Corporation," August 2019 (Business Roundtable, 2019)

★ ★ ★

Unnerved by fundamental economic changes and the failure of government to provide lasting solutions, society is increasingly looking to companies, both public and private, to address pressing social and economic issues. These issues range from protecting the environment to retirement to gender and racial inequality, among others […]

Purpose is not a mere tagline or marketing campaign; it is a company's fundamental reason for being – what it does every day to create value for its stakeholders. Purpose is not the sole pursuit of profits but the animating force for achieving them. Profits are in no way inconsistent with purpose – in fact, profits and purpose are inextricably linked. Profits are essential if a company is to effectively serve all of its stakeholders over time – not only shareholders, but also employees, customers, and communities. Similarly, when a company truly understands and expresses its purpose, it functions with the focus and strategic discipline that drive long-term profitability. Purpose unifies management, employees, and communities. It drives ethical behavior and creates an essential check on actions that go against the best interests of stakeholders. Purpose guides culture, provides a framework for consistent decision-making, and, ultimately, helps sustain long-term financial returns for the shareholders of your company.

Larry Fink, Letter to CEOs, January 2019 (Fink, 2019)

★ ★ ★

A purposeful business will organise itself on all levels according to its purpose. We propose eight principles for business leaders and policymakers. They do not prescribe specific actions, but set out the features of an operating environment that will enable the delivery of those purposes, while remaining flexible to a diversity of business models, cultures and jurisdictions.

1. Corporate law should place purpose at the heart of the corporation and require directors to state their purposes and demonstrate commitment to them.
2. Regulation should expect particularly high duties of engagement, loyalty and care on the part of directors of companies to public interests where they perform important public functions.
3. Ownership should recognise obligations of shareholders and engage them in supporting corporate purposes as well as in their rights to derive financial benefit.
4. Corporate governance should align managerial interests with companies' purposes and establish accountability to a range of stakeholders through appropriate board structures. They should determine a set of values necessary to deliver purpose, embedded in their company culture.

5. Measurement should recognise impacts and investment by companies in their workers, societies and natural assets both within and outside the firm.
6. Performance should be measured against fulfilment of corporate purposes and profits measured net of the costs of achieving them.
7. Corporate financing should be of a form and duration that allows companies to fund more engaged and long-term investment in their purposes.
8. Corporate investment should be made in partnership with private, public and not-for-profit organisations that contribute towards the fulfilment of corporate purposes.

British Academy, Principles for Purposeful Business, November 2019 (The British Academy, 2019, pp. 8–9)

★ ★ ★

a. The purpose of a company is to engage all its stakeholders in shared and sustained value creation. In creating such value, a company serves not only its shareholders, but all its stakeholders – employees, customers, suppliers, local communities and society at large. The best way to understand and harmonize the divergent interests of all stakeholders is through a shared commitment to policies and decisions that strengthen the long-term prosperity of a company […]
b. A company is more than an economic unit generating wealth. It fulfils human and societal aspirations as part of the broader social system. Performance must be measured not only on the return to shareholders, but also on how it achieves its environmental, social and good governance objectives. Executive remuneration should reflect stakeholder responsibility.
c. A company that has a multinational scope of activities not only serves all those stakeholders who are directly engaged, but acts itself as a stakeholder – together with governments and civil society – of our global future. Corporate global citizenship requires a company to harness its core competencies, its entrepreneurship, skills and relevant resources in collaborative efforts with other companies and stakeholders to improve the state of the world.

World Economic Forum, Davos Manifesto 2020, December 2019 (Schwab, 2019)

★ ★ ★

In order to secure our future and to prosper, we need to evolve our economic model. Having been engaged in these issues since I suppose 1968, when I made my first speech on the environment, and having talked to countless experts across the globe over those decades, I have come to realise that it is not a lack of capital that is holding us back, but rather the way in which we deploy it. Therefore, to move forward, we need nothing short of a paradigm shift, one that inspires action at revolutionary levels and pace [...]

Sustainable markets generate long-term value through the balance of natural, social, human and financial capital. Systems-level change within sustainable markets is driven by consumer and investor demand, access to sustainable alternatives and an enhanced partnership between the public, private and philanthropic sectors. Sustainable markets can also inspire the technology, innovation and scale that we so urgently need.

HRH The Prince of Wales at the World Economic Forum in Davos, Switzerland, January 2020 (Prince Charles, 2020)

★ ★ ★

In a free enterprise, the community is not just another stakeholder in the business but is in fact the very purpose of its existence.

Jamsetji Tata, Founder of Tata Group, quotes in India Times, September 2019 (Somvanshi, 2019)

★ ★ ★

Business leaders seem to have found a unified voice to express their view on the purpose of business and more specifically the fact that the purpose of business comes from its being embedded into an ecosystem of actors. The relationships between companies and the constellation of stakeholders that make up these systems are varied and reciprocal: businesses serve their stakeholders and vice versa in many different and essential ways. Stakeholders enter into relationships in order to achieve some bigger purpose beyond their own continued existence, usually tied to some good in the world.

Common yet implicit to all of these statements is a conception of morality, ethical behavior, and the greater good companies are supposed to serve. Inquiry into these questions is in itself very valuable and even important to understand the cultural, philosophical, ideological, and social backgrounds within which these visions of the purpose of business emerge. However, it would take us much too far away from our current goals to examine the vast theoretical landscape undergirding this aspect of the issue. For our current objectives, it is sufficient to note, first, that business is adopting a broader view than the fulfillment of "business interest" in order to explain the transformation that is needed and second, that the continued existence of our society under climate conditions where we can thrive is part of the good that is sought.

Concerning the broader view, this implies that what these statements are calling for will draw on other spheres of society than can be regulated from the boardroom. What "good" ought to direct our actions and organize our society is more likely going to happen in many other fora of the public sphere: educational systems, the media, families, religious communities, and anywhere where people gather and define their common good. This goes far beyond the scope of this book. This book is firmly within the scope of contributing to the dynamic exchanges that make up our common project of defining our ethical orientations, recognizing that decisions in boardrooms and factory floors also feed into these essential questions. This book rests on the assumption that transforming the economy to avert the worst impacts of climate change is an answer to these questions, an assumption this book aims to justify throughout.

Another thread common to all of the foregoing statements is that nothing short of a deep transformation of the way we do business – and by implication, of the economy as a whole – will do for achieving their vision of a future where we thrive and where business can fulfill its purpose. It relies on redesigning the web of relationships with stakeholders and even redefining who those stakeholders are. It also requires including criteria for business success that will be beyond profitability exclusively; they will be compatible with financial sustainability and for some of these views even with growth. They will call into question what has been taken as corporate orthodoxy and offer an opportunity to rethink the place of persons and of nature in business. In this chapter, we will focus on the place of persons in the new corporate climate leadership within the background of rethinking the purpose of business.

These visions also all understand the role of the organization in relation to all of its stakeholders, understanding that cultivating communities, consumers, and employees as well as shareholders, who currently hold the primacy of corporate relations, is a key to a company's purpose and hence to its success. The interests of all stakeholders are, in a manner of speaking, in the interest of the company: not simply so that it can continue to exploit those relationships for financial gain, but so that the value it creates (financial and extra-financial) is beneficial to all stakeholders.

Finally, all these views share an underlying vision that this purpose can be served only by an organization that is designed to do so. Governance geared toward creating the operating environments that allow for purpose to be a goal of business decision-making and replaces the notion of a web of stakeholders, people, and nature at the center of it is needed to bring about the kinds of changes these statements put forth. This governance will necessarily be different than the kind of governance that puts profit and growth at the helm of business goals.

The above statements are representative of this shift in vision and offer some indication as to what purpose consists in and what kind of governance is likely to bring it about. We saw in Chapter 3 a brief definition of the purpose of a company. Purpose is defined extra-financially indicators and expresses what is authentic and unique to a company by articulating the need it fulfills in the world. Understood as a company's "north star," it directs all of its actions in a way that serves the systems and stakeholders (i.e. the world) in which a company operates. Indeed, it is not a mere tagline or catchphrase but an organizing principle that will have a ripple effect on all activities of a company and the entirety of its operating environment. Notably, it will be able to have that effect thanks to a governance that is organized to serve this purpose. What does this mean for companies? What does this mean for climate action? In the rest of the chapter, we will lay some ground showing how purpose can shape corporate governance in general before exploring how this can accelerate climate action.

Purpose and corporate governance

The contribution of purpose is something each company works out and expresses in very concrete steps and decisions. This relies on a mapping of that company's operations, culture, and operating environment and changes resulting from an alignment between purpose and those aspects of the company. It supersedes mission and values and is an overarching "north star" that helps define these. Purpose alone does not guarantee that a company will tackle systemic issues in general and the climate crisis in general. It must be reflected and carried forward by a governance that allows for purpose-driven decisions to be made at all levels. As the statements that open this chapter show, on the one hand that purpose colors all dimensions of a company's activities and on the other it is necessary in order to adapt to a changing world in the present so as to create a world we want in the future.

Corporate governance, broadly speaking, is the rules framework that allows the purveyors of finance to control how that finance is used and usually involves ways to guarantee a return on investment for themselves (Segrestin, Levillain and Hatchuel, 2016). This conception has evolved to define "governance as the determination of the broad uses to which organizational resources will be deployed and the resolution of conflicts among the myriad participants in organizations" (Daily, Dalton and Cannella Jr., 2003, p. 371). There are various theories and much literature on the topic of governance, known as "agency theory," "stewardship theory," and "stakeholder theory." What is important to note is that they are all premised on the notion of monitoring and controlling the management of financial resources so as to maximize growth and profits. The difference in these theories tends to focus on what falls under the remit of what is monitored and controlled but the objective remains the same. In other words, they are all concerned with the same "why," but have different views on the "how" (Alchian and Demsetz, 1972, pp. 57–74; Clarkson, 1995, pp. 92–117; Donaldson and Davis, 1991, pp. 49–64; Donaldson and Preston, 1995, pp. 65–91; Eisenhardt, 1989, pp. 57–74; Gabrielsson, Huse and Minichilli, 2007, pp. 21–39; Hernandez, 2012, pp. 172–193; Jensen and Meckling, 1976, pp. 1–35; Mitchell, Agle and Wood, 1997, pp. 853–886; Moriarty, 2014, pp. 31–53; Tirole, 2001, pp. 1–35).

More precisely, the rules of corporate governance, in general, affect the internal and external actions of employees as well as relationships with external stakeholders. In addition, they specify outline duties, privileges, and roles of board members and executives to ensure that they behave in the best interest of the company. Corporate governance typically also includes directives about the role of shareholders, and in this aspect of governance we often see specifications on disclosure, dividends, and shareholder activism. It also sets out rules about the goals and objectives of business contracts such as the rate of return, contract length, and contract approval. The system of governance further functions as a system of checks and balances for internal business departments, being usually designed so as to ensure that departments do not unduly dominate the business or operate outside the company's mission or values. Corporate governance is essential to a company not only since it provides a framework of regulations and rules but also because it leaves room to develop tailored practices that suit the company's needs and to adapt those in light of new conditions or standards.

Given these functions of corporate governance as they are traditionally understood, purpose-driven corporate governance distinguishes itself by serving a different "why" altogether. Going back to the opening statements and using this lens, it is striking that the "why" of companies is presented as including but also going beyond "delivering value" for stakeholders. On those views, governance is designed so as to generate investment in employees or communities, the ethical and transparent management of relationships, long-term strategic planning. What makes these special is the nature of the value that is delivered: while it might include profits and returns for shareholders on certain of these visions, it is not reduced to that. Indeed, value here is value for the future, value for communities and people understood not only as resources but as entities that deserve respect and may hope to thrive in the world, value for "human and societal aspirations as part of the broader social system," value for our "global future" via "systems-level change."

Consequently, we can expect purpose-driven corporate governance to differ from more traditional versions in that it will be designed so as to make sure that this kind of value is delivered. Concretely, this means that the way relationships are managed will be geared toward purpose, the duties of board members will be too. Contracts, supplier relations, risk, and financial disclosures will all be affected. Table 8.1, which is not exhaustive, gives some examples of the kinds of changes that would be entailed by adopting corporate-driven governance.

Table 8.1 is illustrative and not exhaustive. Indeed, a full table can be drawn only for a given company with a specific and unique purpose and profile. What is more, there are new systems, practices, and measures that remain to be invented and conceived, and we do not wish to constrain the imagination of leaders in this area. However, a few common principles can nonetheless help guide the design of purpose-driven corporate governance: a longer time horizon, the centrality of the person and not merely the employee or the supplier, transparency and consistency as guiding financial activities, and equality, fairness, and accountability being embedded into decision-making processes. Each company will determine

TABLE 8.1 Elements of purpose-driven corporate governance

Area of operations	Measures to implement	How it advances purpose
Financial disclosure	Risk disclosure, climate disclosure via TCFD	Better knowledge and management of social and environmental footprint of a company
Supplier relations	Long-term relationship management via, e.g. trainings, sustainable financing, participation in decision-making	Brings prosperity and welfare on fair and egalitarian terms to communities and suppliers
Supplier relations	Due diligence on third party relations to monitor risk (Refinitiv, 2020)	Gain insight on and mitigate third-party risk, including the risk of unethical and unfair practices and relationships
Human resources and labor relations	Strengthen employee representation and participation in boards	Bring about fairer and more egalitarian decision-making processes, more accountability in decision-making
Fiscal policy	Abolish offshore accounts in tax havens	Commit to full transparency so as to monitor investment in toxic industries and favor fair fiscal contributions in countries of operation and employment
Government and public affairs	Streamline lobbying activities to ensure consistency and avoid contradictory efforts	Stop sending mixed messages to policy makers and other stakeholders; higher likelihood of reaching goals
Corporate financing	Introduce forms and durations of financing that allow companies to engage in long-term and purpose-driven projects	Rule out short-term pressure for financial performance and favor instead purpose-driven activities and goals
Board duties and responsibilities	Elevate purpose-related issues such as climate or gender fairness at the board level	Ensure accountability, transparency on core purpose related issues and elevate them to the level of duty equal to profit and growth

what new form of corporate governance best serves its purpose and will implement it in a way that is aligned with it.

Corporate governance and climate leadership

So far, we have seen a growing and influential trend that companies are expected to be purpose-driven in order to be adapted to the future we want and even to be able to lead is into it. What is more, taken seriously, purpose has deep impacts on a company insofar as it determines corporate governance. What does this have to do with climate action and climate leadership?

One of the main arguments put forward by this book is that climate leadership calls for a rethinking of deeply rooted assumptions about companies, their role, and the economic system they are in a reciprocal relationship with, being enabled by it and supporting it themselves. One of these beliefs is that climate action is the exclusive remit of environmental measures with an eye, in the best of cases, on social impacts. This vision, we argue, is not sufficient to tackle the task at hand, which is one of unprecedented scale that requires systems change at a rapid pace. Environmental measures will be crucial, and this transformation will not be achieved without massive emissions reductions in all global value chains and reaching net-zero emissions by 2050. However, even this change will require an enabling environment for long-term investments and deep changes in current value chains. In addition, this change, necessary as it is, will not be sufficient to avert the worst impacts of climate change through adaptation measures, nor will it be suited to ushering in a new economic system where a thriving nature and society becomes the priority value to be protected and shared.

Does purpose necessarily involve a climate dimension? Given the wide diversity of companies, sectors, geographies, all with their unique web of stakeholders and with their own histories, one might expect that there is no single answer to this question. However, purpose, as it is envisioned by business leaders, involves a perspective on the future, on the natural world, and on universal social goals and aspirations, which are only possible under conditions of more equality and relative prosperity. Hence there can be no purpose as it is understood by corporate practitioners without regard for the climate emergency.

An effective climate governance structure is critical to ensure that a company is properly equipped to tackle climate change. Indeed, as a growing number of companies set emissions targets aligned with science, these targets are even becoming a part of the set of indicators that demonstrate the robustness of a company's risk management strategy. Moreover, regardless of whether climate impacts are specified as a fiduciary duty under the corporate governance code of a company, directors have the duty to promote the success of the company and to act with due care, skill, and diligence. If climate risk is material to a company, failure to identify, assess, deal with or disclose amounts to a failure of corporate governance.

What emerges from the foregoing is that there are at two distinct ways in which the transformation of corporate governance is a pillar of the transformation of the economy in order to address climate change, which we will explore in turn: (1) how purpose-driven corporate governance enables strategic risk management and availing of opportunities related to climate action, and (2) how purpose-driven corporate governance can help companies contribute to the transformation of the economy toward low emissions or net zero.

Corporate governance and risk management, and climate opportunities

We saw that climate change is a potential strategic risk to companies in Chapter 1, with exposure to these risks being particularly high companies that operate in sectors that are more vulnerable to climate risks, namely energy, transportation, agriculture, food and forest products, materials and buildings, and financial services. It ought to be the duty of the board of directors in all sectors to identify and manage it in the same way as any other strategic risk. We further saw in Chapter 2 that companies stand to reap multiple benefits from climate action, advancing many core business objectives. A purpose-driven company will aim to design its governance to match its duty to protect itself from this kind of risk and to maximize the likelihood it will avail of the right opportunities.

Where risk is concerned, once a company has integrated climate risk as part of its set of material risks, directors will be expected to take strategic decisions on how to manage those risks and their associated opportunities. This includes implementing measures such as:

- Undertaking appropriate risk assessments in order to understand how a company or group of companies is projected to be affected by the different dimensions of climate risk.
- Connecting risk mapping and management at the executive level to risk oversight at the board level in order to create accountability.
- Developing scenario-planning and forecasting of strategic risks and opportunities related to low-carbon transition, including for instance carbon pricing, emissions reductions targets, fixed assets, capital allocation, R&D, in order better to plan for the future of the company.
- Developing a robust mapping of financial exposure to climate risk that will impact cost structure, regulatory costs, capital spending, cash flow, insurance, in order to secure the financial sustainability of the company.
- Adjusting investment and financing standards to allow for longer-term rates of return.
- Disclosing material climate risk in order to increase transparency about exposure and measures taken to mitigate risk.
- Roll out targets and metrics in alignment with risk mapping.
- Disclosing about corporate governance around climate in order to demonstrate board accountability to stakeholders.

With this kind of risk management, companies will be held accountable for climate risk in much the same way that they are accountable for enterprise risk of any other kind. Different business units and different departments will be empowered to take measures that directly relate to this aspect of corporate governance. Specifically, they too will be tasked with mitigating climate risk via, for instance, aggressive emissions reductions or ambitious adaptive measures. On the other hand, actions that allow the company to avail of climate opportunities via investment, operational changes, supply chain management changes will possibly be vetted and supported (Breitinger et al., 2018).

By updating its purpose and making governance and strategy decisions accordingly Mercedes Benz and its parent company Daimler decided in 2019 to stop manufacturing internal combustion engines in order eventually to offer a completely electric fleet. Mercedes's website states that

> The Purpose of Mercedes-Benz Cars has been there since Gottlieb Daimler and Carl Benz invented the automobile – but it was only officially formulated in spring 2018. To this end, several hundred interviews were conducted worldwide from summer 2017, over 200 historical and current documents were evaluated and more than 23,000 employees from different regions, functions and hierarchical levels were surveyed… "First Move the World" means pursuing more than what is immediately achievable. This pioneering spirit is part of our DNA. This is why it also gives guidance for the all-embracing and sustainable Mercedes-Benz Cars business strategy, and the basis for our strategic decisions. It gave rise to, for example, Ambition 2039, our road to sustainable mobility, green production with CO_2-neutral energy supply and the long-term development cooperation for automated driving with Daimler AG and the BMW Group. (Mercedez-Benz, 2019)

Drawing from history and a rich web of stakeholders, the companies aimed to uncover the need they served in the world and aligned their governance and strategy to it. As a result, Daimler is once again leading the sector into a new phase of its existence, this one away from fossil fuel energy and toward the decarbonized economy we need. Availing of opportunities – including the opportunity to position themselves as leaders – Daimler and Mercedes Benz are also protecting themselves against the risks of deep transformation looming over the auto industry.

By embedding tackling the climate emergency within its purpose, a company can protect itself from the various risks related to climate change that are looming in all sectors. Purpose, and ensuing changes to governance with their own ripple effect of changes downstream in a company's operations, being unique to a company will help a business translate the economy-wide risks into an equally unique risk mitigation strategy and into unique opportunities tied to decarbonization.

Corporate governance and the transformation of the economy toward low emissions or net zero

Purpose-driven corporate governance can certainly help companies reduce their emissions and decarbonize their value chains like Daimler did. Corporate governance can, as we have seen, favor direct climate action in various divisions of a company. However, we have also seen in Chapter 3, for instance, that emissions reductions that are part of an unchanging context of the exclusive pursuit of growth will not suffice to transform the economy toward net zero.

While enabling direct climate action, such as the setting of ambitions emissions targets, risk disclosure and reporting, and long-term investment in R&D,

is necessary to this transformation, it is not sufficient. What is needed is the cultivation of an entire ecosystem that moves businesses away from operational practices that block climate action and climate justice. In practice, this means looking beyond the typical remit of climate-related governance to other areas: transparency, fiscal accountability, forms of participatory decision-making and labor relations, lobbying practices more broadly. Let us examine these in turn.

Transparency

There is no way to trace capital that is not submitted to transparent processes, when it is, for instance, channeled through tax havens. In this way, companies cannot be held accountable and targets and metrics cannot be verified. Consider, for instance, that between 2000 and 2011, almost 70% – or $18.4 billion of a total $26.9 billion – of foreign capital invested by major companies in soy and beef farming in Brazil had flowed through tax havens. These sectors are responsible for vast amounts of deforestation, and hence these investments were acting as indirect subsidies for environmentally damaging practices (Galaz et al., 2018, pp. 387–404). Fishing and fisheries are also some of the most affected sectors by lack of transparency.

Fiscal accountability

Preparing for the biggest impacts of climate change, e.g. to respond to natural disasters or population displacements, or investing into innovation for the transition to low-carbon modes of production, transport, and consumption, requires public investment on scales rarely deployed. Government revenue is also central to the resilience of its population via investments in health, education, and political equality. Fiscal accountability, including paying taxes, is a building block of the transition toward a net-zero economy.

All of these investments require that governments be able to collect taxes reliably and fairly, especially corporate taxes. Developing economies in particular rely heavily on corporate income tax bases. The IMF reports that the revenue lost from tax evasion in these countries is 1.3 times larger, as a share of GDP, than it is in advanced economies (Crivelli, De Mooij and Keen, 2015, p. 21). And this is in a context where corporate tax rates are low, to begin with. At an average of 17.6% of GDP, tax collection in the Asia-Pacific region for instance is among the lowest in the world, a level regarded by experts as totally insufficient to achieve sustainable development goals, and by implication, a net-zero economy (Isgut and Jian, 2017, p. 3).

Participatory decision-making and labor relations

Trade unions have a unique role in raising awareness about climate and pushing for ambitious climate action. For instance, they can build a community for climate action, engage in workplace action to drive energy efficiency, recycling, and other environmental reforms, create solidarity and direct action in support of communities facing climate catastrophes, and partner with climate and development groups to mobilize for political commitment. They are also key in negotiations about industrial change, job loss, and green jobs. Indeed, case studies of such actions abound (ILO, 1999).

The role of labor is broader than direct climate action by affecting the entire labor process in ways that favor not only a shift to net zero but a shift away from the hegemony of exclusive growth-seeking. The role of unions in particular in negotiations around international trade has ripple effects in resisting this kind of market logic. In their opposition to NAFTA, for instance, US labor unions were calling for specific decisions but also for the overall protection of workers' rights and their prioritization over the kind of profit-seeking NAFTA allowed, which at the same time relied on the erosion of environmental protection, human rights, and worker's rights (Prentice and Lawder, 2019). Unions are not only fighting for discreet items that are favorable to climate action but also to rethink the very purpose of companies and their trade relations and how all of these shape the economy and the world.

Now considering that in the 2015–2016 election cycle, business outspent unions 16-to-1 ($3.4bn versus $213m) and that each year US unions spend ca. $48m on lobbying, while corporate America spends more than $2.5bn – more than 50 times as much. Consider too that company anti-union practices are common, with, for instance, 81% of countries having violated the right to collective bargaining in 2018 with the support or tolerance of the private sector (ITUC, 2018, p. 6). The weakening of unions and the outright hostility demonstrated by outspending signals that there is ample room for companies to change their labor relations in ways that will be favorable to reaching the transformation of the economy the climate emergency calls for.

Lobbying

The five largest publicly traded oil and gas majors – ExxonMobil, Royal Dutch Shell, Chevron, BP, and Total – invested over $1bn in the three years following the Paris Agreement on misleading climate-related branding and lobbying (Influence Map, 2019). These companies and others in their sector do not have, historically, a reputation for transparency. Indeed, it is documented that they have for decades aimed to stop the fight against climate change by using a variety of tactics, some overt and some not. In 1989, one year after the establishment of the IPCC, the largest oil and gas and coal firms came together to launch the Global Climate Coalition (GCC), a group dedicated to spreading doubt and skepticism about the scientific basis of climate change (Brown, 2000; Revkin, 2009). All the while, documents have emerged showing that these companies' scientists and executives were aware of the soundness of the science, not to speak of the urgency of the matter; indeed sometimes it emerged from their own research (Climate Investigations Center, 2021; *The New York Times*, 2009). In light of this history and of the scale of lobbying spend, statements in support of global climate action and about the constructive role of the industry are marred with a lack of credibility (ExxonMobil, 2016).

The main issue with lobbying against climate action is not the impact on the reputation or credibility of these companies. What matters most is that they actually block climate action by providing means and resources to regressive organizations and causes. As these companies start seeing the inevitability of the transformation of the economy, and for some the various opportunities associated with it, consistency in their actions will contribute to accelerating its pace rather than blocking it or slowing it down as they have for the last half-century.

Companies' actions beyond the strict perimeter of environmental and climate action must be scrutinized and aligned with the climate emergency. The scope of climate action has typically been restricted to mitigation efforts and sometimes to adaptation efforts, overlooking the broader context that must be mobilized for the transformation of the economy: finance, capital, political influence, and labor relations. Changes in these areas are enabled by a corporate governance that is oriented toward this transformation and that has integrated it, as a forward-looking, people-centered, and planet-respecting necessity, into its purpose.

It is more likely that companies that do this will seek to deliver genuine value to customers, invest in their employees, deal fairly and ethically with suppliers, and support the communities in which they derive the social license to operate. This reflects a growing understanding that business success depends on purpose, which, broken down, means shaping corporate governance so as to secure operations, mitigate risk, build a strong reputation, recruit and retain the best talent. It is plainly not sufficient to reduce emissions in order to shift the economy in the direction it needs to go in order to avert the worst impacts of climate change – nothing less than the deep transformation of corporate governance in line with purpose and hence of relationships with complex systems of stakeholders will do.

References

Alchian, A.A. and Demsetz, H. (1972) 'Production, Information Costs and Economic Organization,' *The American Economic Review*, 62(5) [Online]. Available at: https://www.jstor.org/stable/1815199?seq=1 (Accessed: 15 November 2020)

Breitinger, D. et al. (2018) *How to Set Up Effective Climate Governance on Corporate Boards Guiding Principles and Questions*. Available at: http://www3.weforum.org/docs/WEF_Creating_effective_climate_governance_on_corporate_boards.pdf (Accessed: 20 March 2021)

Brown, L.R. (2000) 'The Rise and Fall of the Global Climate Coalition,' *Earth Policy Institute*, 25 July [Online]. Available at: http://www.earth-policy.org/plan_b_updates/2000/alert6 (Accessed: 20 March 2021)

Business Roundtable (2019) Our Commitment [Online]. Business Roundtable Opportunity. Available at: https://opportunity.businessroundtable.org/ourcommitment/ (Accessed: 20 March 2021)

Clarkson, M.E. (1995) 'A Stakeholder Framework for Analysing and Evaluating Corporate Social Performance,' *Academy of Management Review*, 20(1) [Online]. Available at: https://www.jstor.org/stable/258888?seq=1#metadata_info_tab_contents (Accessed: 15 November 2020)

Climate Investigations Center (2021) Exxon Documents [Online]. Climate Investigations Center. Available at: https://climateinvestigations.org/exxonknew/ (Accessed: 20 March 2021)

Crivelli, E., De Mooij, R. and Keen, M. (2015) *Base Erosion, Profit Shifting and Developing Countries*. IMF Working Paper No 15-118. Available at: https://www.imf.org/external/pubs/ft/wp/2015/wp15118.pdf (Accessed: 20 March 2021)

Daily, C.M., Dalton, D.R. and Cannella Jr, A.A. (2003) 'Corporate Governance: Decades of Dialogue and Data,' *Academy of Management Review*, 28(3) [Online]. Available at: https://www.jstor.org/stable/30040727?seq=1 (Accessed: 15 November 2020)

Donaldson, L. and Davis, J.H. (1991) 'Stewardship Theory or Agency Theory: CEO Governance and Shareholder Returns,' *Australian Journal of Management*, 16(1) [Online]. Available at: https://journals.sagepub.com/doi/10.1177/031289629101600103#:~:text=Agency%20theory%20argues%20that%20shareholder,shared%20incumbency%20of%20these%20r%C3%B4les (Accessed: 15 November 2020)

Donaldson, L. and Preston, L.E. (1995) 'The Stakeholder Theory of the Corporation: Concepts, Evidence and Implications,' *Academy of Management Review*, 20(1) [Online]. Available at: https://www.jstor.org/stable/258887?seq=1#metadata_info_tab_contents (Accessed: 15 November 2020)

Eisenhardt, K.M. (1989) 'Agency Theory: An Assessment and Review,' *Academy of Management Review*, 14(1) [Online]. Available at: https://www.jstor.org/stable/258191?seq=1#metadata_info_tab_contents (Accessed: 15 November 2020)

ExxonMobil (2016) 'Statements on Paris Climate Agreement,' *ExxonMobil*, 4 November [Online]. Available at: https://corporate.exxonmobil.com/Sustainability/Environmental-protection/Climate-change/Statements-on-Paris-climate-agreement#ExxonMobilstatementonCOP21 (Accessed: 20 March 2021)

Fink, L. (2019) *Larry Fink's 2019 Letter to CEOs. Purpose & Profit* [Online]. BlackRock. Available at: https://www.blackrock.com/corporate/investor-relations/2019-larry-fink-ceo-letter (Accessed: 20 March 2021)

Gabrielsson, J., Huse, M. and Minichilli, A. (2007) 'Understanding the Leadership Role of the Board Chairperson Through a Team Production Approach,' *International Journal of Leadership Studies*, 3(1) [Online]. Available at: https://www.researchgate.net/publication/230690276_Understanding_the_Leadership_Role_of_the_Board_Chairperson_through_a_Team_Production_Approach (Accessed: 15 November 2020)

Galaz et al. (2018) 'Tax Havens and Global Environmental Degradation,' *Nature Ecology & Evolution*, 2 [Online]. Available at: https://www.nature.com/articles/s41559-018-0497-3 (Accessed: 15 November 2020)

Hernandez, M. (2012) 'Toward an Understanding of the Psychology of Stewardship,' *Academy of Management Review*, 37(2) [Online]. Available at: https://www.researchgate.net/publication/265979974_Toward_an_Understanding_of_the_Psychology_of_Stewardship (Accessed: 15 November 2020)

ILO (1999) *Trade Union Actions to Promote Environmentally Sustainable Development*. Available at: https://www.ilo.org/wcmsp5/groups/public/—ed_dialogue/—actrav/documents/publication/wcms_122116.pdf (Accessed: 20 March 2021)

Influence Map (2019) *Big Oil's Real Agenda on Climate Change*. Available at: https://influencemap.org/report/How-Big-Oil-Continues-to-Oppose-the-Paris-Agreement-38212275958aa21196dae3b76220bddc (Accessed: 20 March 2021)

Isgut, A. and Jian, Z. (2017) *Issues Paper on Tax Policy and Public Expenditure Management in Asia and the Pacific*. United Nations ESCAP Working Paper No 17-06. Available at: https://www.unescap.org/sites/default/files/publications/WP-17-06_Issues%20Paper%20on%20Tax%20Policy_0.pdf (Accessed: 20 March 2021)

ITUC (2018) *2018 ITUC Global Rights Index. The World's Worst Countries for Workers*. Available at: https://www.ituc-csi.org/ituc-global-rights-index-2018 (Accessed: 20 March 2021)

Jensen, M.C. and Meckling, W.H. (1976) 'Theory of the Firm: Managerial Behavior, Agency Costs and Ownership Structure,' *Journal of Financial Economics*, 3(4) [Online]. Available at: https://www.sciencedirect.com/science/article/pii/0304405X7690026X (Accessed: 15 November 2020)

Mercedes-Benz (2019) '*First Move the World' The Purpose of Mercedez-Benz Cars* [Online]. Mercedes-Benz. Available at: https://www.mercedes-benz.com/en/company/purpose/ (Accessed: 20 March 2021)

Mitchell, R.K., Agle, B.R. and Wood, D.J. (1997) 'Toward a Theory of Stakeholder Identification and Salience: Defining the Principle of Who and What Really Counts,' *Academy of Management Review*, 22(4) [Online]. Available at: https://www.jstor.org/stable/259247?seq=1#metadata_info_tab_contents (Accessed: 15 November 2020)

Moriarty, J. (2014) 'Compensation Ethics and Organizational Commitment,' *Business Ethics Quarterly*, 24(1) [Online]. Available at: https://www.cambridge.org/core/journals/business-ethics-quarterly/article/abs/compensation-ethics-and-organizational-commitment/294A453366E268A2F3856D4FA96AD22C (Accessed: 15 November 2020)

Prentice, C. and Lawder, D. (2019) 'U.S. Labor Unions Say NAFTA Replacement Does Not Go Far Enough for Workers,' *Reuters*, 26 March [Online]. Available at: https://www.reuters.com/article/us-usa-trade-labor/us-labor-unions-say-nafta-replacement-does-not-go-far-enough-for-workers-idUSKCN1R72H2 (Accessed: 20 March 2021)

Prince Charles (2020) 'We Need Revolutionary Action to Save the Planet: Full Transcript of Prince Charles' Davos Speech,'
The Sydney Morning Herald, 23 January [Online]. Available at: https://www.smh.com.au/environment/climate-change/we-need-revolutionary-action-to-save-the-planet-full-transcript-of-prince-charles-davos-speech-20200123-p53tyl.html (Accessed: 20 March 2021)

Refinitiv (2020) *The Real Risks: Hidden Threats within Third-Party Relationships*. Available at: https://www.refinitiv.com/en/risk-and-compliance/resources/hidden-threats-third-party-risk (Accessed: 20 March 2021)

Revkin, A.C. (2009) 'Industry Ignored Its Scientists on Climate,' *The New York Times*, 23 April [Online]. Available at: https://www.nytimes.com/2009/04/24/science/earth/24deny.html (Accessed: 20 March 2021)

Schwab, K. (2019) 'Davos Manifesto 2020: The Universal Purpose of a Company in the Fourth Industrial Revolution,' *World Economic Forum*, 2 December [Online]. Available at: https://www.weforum.org/agenda/2019/12/davos-manifesto-2020-the-universal-purpose-of-a-company-in-the-fourth-industrial-revolution/ (Accessed: 20 March 2021)

Segrestin, B., Levillain, K. and Hatchuel, A. (2016) 'Purpose-Driven Corporations: How Corporate Law Reorders the Field of Corporate Governance,' European Academy of Management Conference (EURAM 2016), Paris, France [Online]. Available at: https://hal.archives-ouvertes.fr/hal-01323118/document (Accessed: 15 November 20)

Somvanshi, K.K. (2019) 'India Inc Gradually Finds a Purpose Beyond Profit,' *The Economic Times*, 9 September [Online]. Available at: https://economictimes.indiatimes.com/news/company/corporate-trends/india-inc-gradually-finds-a-purpose-beyond-profit/articleshow/71041180.cms?utm_source=contentofinterest&utm_medium=text&utm_campaign=cppst (Accessed: 20 March 2021)

The British Academy (2019) *Principles for Purposeful Business.* Available at: https://www.thebritishacademy.ac.uk/documents/224/future-of-the-corporation-principles-purposeful-business.pdf (Accessed: 20 March 2021)

Tirole, J. (2001) 'Corporate Governance,' *Econometrica*, 69(1) [Online]. Available at: https://onlinelibrary.wiley.com/doi/10.1111/1468-0262.00177 (Accessed: 15 November 2020)

The New York Times (2009) *Advisers to Industry Group Weigh in on Warming* [Online]. *The New York Times.* Available at: https://www.nytimes.com/interactive/projects/documents/global-climate-coalition-aiam-climate-change-primer#p=1 (Accessed: 20 March 2021)

9
FROM EVOLUTION TO REVOLUTION

Civil society and rethinking business in a new climate economy

Climate change is a totalizing issue of our epoch, an issue with which any future-oriented collective action must contend. The Paris moment has set in motion something we have never seen before – an international agreement rooted in diplomacy that gave us a glimpse into the foundations of Kantian perpetual peace.

Corporate leadership cannot ignore forces outside itself, in particular forces coming from civil society. Throughout this book, we have presented the architecture of the web of stakeholders that make up the system within which companies can lead, the ways in which this web exerts pressure on business, and the ways in which businesses can respond to these pressures. Civil society makes up a significant part of this web, and increasingly business has entered into dialogue with it, sometimes constructively, sometimes reactively. Most companies have historically focused on their core business – their own operations and their supply chain – and avoided exposing themselves to the scrutiny of civil society. On the other hand, civil society has a long tradition of holding companies to account for the impact they have on the societies in which they operate. As this society becomes increasingly global and transparent, companies can no longer avoid nor control this exposure fully. This chapter takes a step back in looking at how civil society's role in shaping major evolutions in corporate practices and even in the economy at large. Secondly, it explores the role of civil society with regard to climate action in particular with a view to understanding how business can best engage with actors within it. Any company considering taking on a leadership role in the transition to the new climate economy will need to understand the dynamics and mechanisms of civil society within which it operates and will have a significant role in framing the entire field of climate action.

Civil society in context

The World Bank tells us that civil society includes a wide array of organizations: "non-governmental organizations, community groups, labor unions, indigenous peoples movements, faith-based organizations, professional associations, foundations, think tanks, charitable organizations, and other not-for-profit organizations" (The World Bank, 2017). Broad as it is, this category stands out in contradistinction to government and business and as such is sometimes referred to as "the third sector" (United Nations, 2019). The EU defines civil society as "all forms of social action carried out by individuals or groups who are neither connected to, nor managed by, the State" (European Union, 2010). The African Development Bank defines civil society as the voluntary expression of the interests and aspirations of citizens organized and united by common interests, goals, values, or traditions and mobilized into collective action (AFDB, 2012, p.4).

Civil society comprises people who are not acting in roles within government or business as well as organized groups that are referred to as "civil society organizations" or CSOs. Civil society organizations are groups and networks that vary by size, structure, and platform ranging from international non-governmental organizations (e.g. Oxfam) and mass social movements (e.g. the Arab Spring) to small, local organizations (e.g. Coalition of Jakarta Residents Opposing Water Privatization). They represent a spectrum of actors with a wide range of purposes, constituencies, structures, degrees of organization, functions, size, resource levels, cultural contexts, ideologies, membership, geographical coverage, strategies, and approaches. They are not united thematically at a specific level, though they can be characterized as all being guided by the fight for more justice, more equality, and better representation of typically marginalized voices.

The term gained traction in political and economic debates in the 1980s, especially around the topic of development, when it was used to describe non-state movements in Central and Eastern Europe and in Latin America that were rising against authoritarian regimes (World Economic Forum, 2013). While this definition has evolved, civil society maintains an important role in discussions of development as it provides opportunities to bring communities together for collection action, mobilizing society to articulate demands and voice concerns at local, national, regional, and international levels. These functions, of course, are also essential for climate action, as we shall see in more detail below.

Within this diversity, civil society takes on several roles:

- Providing services, such as providing basic community health care services, running literacy programs
- Campaigning or advocacy, usually lobbying governments or businesses on issues such as indigenous rights, women's rights, the environment
- Serving as watchdog, for instance, monitoring government compliance with various treaties, ethical codes

- Mobilizing citizens, encouraging voter turnout, mobilizing engagement at different levels of governments
- Participation in global government processes, by, for instance, serving on advisory boards, engaging in meetings at United Nations summits

Many – though not all – civil society organizations stand for groups and causes that tend to be overlooked or underserviced by other areas of collective action (government and business), namely the youth, women, and minorities of all kinds (ethnic, racial, religious, or related to sexual orientation). In addition, it is often members of these relatively marginalized groups that have led the fights for more equality, more enfranchisement, and better representation. Most famously, perhaps, Rosa Parks catalyzed a milestone era of the Civil Rights movement – the Montgomery Bus Boycott – that eventually led, 13 months later, to the US Supreme Court ruling that segregation on buses was unconstitutional. What is less known is that Rosa Parks had been an important civil rights activist before her decision not to move to the back of the bus and that her impact as such is often overlooked.[1] The fight for LGBTQ rights was also led by minorities, though memory tends to obfuscate this fact: Marsha P. Johnson, for instance, is but one minority figure central to the gay rights movement that is rarely cited. A transgender rights activist and a figurehead during the Stonewall uprising in 1969 that amplified the until-then timid gay rights movement, Marsha P. Johnson, launched a new era of gay rights activism with the kind of leadership that is representative of the role of marginalized persons in civil society. In peacebuilding, it is becoming increasingly more recognized that women hold key roles in civil society: in Kosovo, in Afghanistan, or in Colombia women are starting to be recognized as vital to the peacebuilding mission, though they are not always rewarded for their actions (Baker and Haug, 2002; Kroc Institute, 2019; World Economic Forum, 2019). More recently, we have seen the spread of the MeToo movement and its unprecedented strides toward justice in the area of gender-based violence, oppression, and discrimination. This movement too was founded and lead by a Black woman, Tarana Burke, and its impacts have been beneficial for women of all ethnicities and races, and sometimes it is even perceived to have been co-opted by white women of more privilege and visibility (Griffin, 2019, pp.556–572; MeToo Movement, 2021; Phipps, 2019, pp.1–25). Black Lives Matter was also founded and is led by three Black women in the United States: Patrisse Cullors, Alicia Garza, and Opal Tometi (Fulton, 2020). These examples are just a few illustrations of the fact that the vibrancy and successes of civil society are often thanks to people whose voices have not been placed front and center of the battles they have fought and won. We shall see below the role of these voices and activists in the field of climate action.

Less organized forms of civil society have emerged as new actors: social movements, online activists, bloggers, influencers, and others (VanDyck, 2017, p.3). Enabled by technologies that have themselves emerged in the last two decades such as the internet and social media, these actors are characterized by connection across space and "virality" and rapid dissemination. The Arab Spring, a series of pro-democracy uprisings that took place in a number of Muslim countries, starting in Tunisia in December 2010 and gaining international traction in Spring 2011 by spreading to Morocco, Syria, Libya, Egypt, and Bahrain, reflected that citizens are becoming more networked and avail of opportunities to express themselves on public platforms. The movement was enabled by the internet and social media platforms such as Facebook and Twitter and eventually was supported by more traditional CSOs – trade unions in Tunisia, Egypt, and Bahrein, for instance (Larbi, 2017). In 2020, in the aftermath of the police killing of George Floyd, teenagers in Nashville who met on Twitter organized a protest march that they claim brought together 20,000 people. Their words capture the power of this new kind of informal activism in civil society: "We were born in the digital age, we can do anything" (Lamb, 2020).

As new actors connect and mobilize spontaneously, they are reshuffling the landscape of civil society by highlighting that more traditional forms of civil society organization are not always fit for the purpose of bringing about the desired change. We observe a gap growing between organized "traditional" civil society and the constituencies they putatively represent. Indeed, some organizations have financial ties leading to conflict of interest: a study of 61 large international NGOs in 13 donor countries found that it is neither poverty nor poor governance that influences NGOs' choice of location, but the location of donors (Koch, 2009, p.24). Yet others have been involved in sordid scandals, including concerning sexual exploitation (Gayle, 2018; Wheeler, 2020). As a result, public distrust not only about the legitimacy but also about the very relevance of CSOs is waning. Edelman points out that in a context where trust in CSOs had been consistently growing over the decade prior to 2016 (the last period for which they have measured data), its rate of growth has significantly decreased (Edelman, 2016). Gallup and Wellcome measured through polling that a third of the global population does not trust NGOs (Younis and Rzepa, 2019). As new actors emerge with new modes of engagement, new tools, new approaches, and new demands, the space for advocacy in civil society can become at once broader and more inclusive, and possibly more effective. What is more, as they continue to prove their power through results, new social movements may raise questions about the need for and relevance of more organized forms of civil society activism. These dynamics are at play in the more specific field of climate action, to which we turn now.

Civil society and climate action

Civil society has played a crucial role in the evolution of climate action across to board: from making it acceptable and even non-negotiable to setting the agenda by broadening the concept of climate action, thereby raising the bar as to what kind of climate action is, in fact, acceptable. Given the overwhelming resistance of business since the discovery of the effects of increasing CO_2 emissions in the atmosphere in the 1970s, and the lack of political will to enact any form of regulation or policy strong enough to manage or stall it, it is fair to say that without scientists, NGOs, or grassroots activists we would not have made progress we have made to date in this field.

The starting point of the history of climate activism could extend as far back as resistance to colonial settlers in the Americas who were keen to exploit resources, both natural and human, for the hauling of wealth six centuries ago. Forced migration, the introduction of new biota, and the establishment of new modes of exploitation and production all had lasting consequences on the colonized societies, who in turn attempted to resist this ecological and environmental disruption (Wallman, Wells and Rivera-Collazo, 2018, p.387). While these roots in resistance and activism are important, more relevant to our current study is the emergence of climate activism in the more recent period since science, business, and policy confronted each other in the late 1980s. Indeed, throughout the development of the scientific field of climate change since the 19th century, sophistication increased and with that the complexity of hypotheses and experiments. Talk about climate change was the province of experts, and the concepts of "global warming," "greenhouse effect," or "climate change" were only timidly introduced into the public sphere at that time, entering politics twenty years after the first climate model, with James Hansen of Nasa's now-famous speech to the US Congress in 1988 (US Senate, Committee on Energy and Natural Resources, 1988). We will take this date as a starting point not of civil society taking a role in climate action but of our analysis of it.

In 1990, the G7 decided to create the Intergovernmental Panel on Climate Change (IPCC), the authoritative scientific body that is still the source of science for climate policy to this day. Significantly for our understanding of the history of climate action and civil society, this decision was not only welcomed but arguably spearheaded by two conservative heads of state, Margaret Thatcher and Richard Nixon (Nulman, 2015). Born out of a political decision to have an independent institution to support sound policy making. Two years later, after receiving the first report of the IPCC, the international community for the first time heeded the alarm signals of the scientific community at the Earth Summit in Rio in 1992. And hence a few years later, in 1995, the first Conference of the Parties took place in Berlin under the leadership of then Environment Minister Angela Merkel, marking the first step of a long process of global climate negotiations within the UNFCCC framework that was established for that very purpose. This provided the stage for one of the first formal confrontations between countries of different stages of industrialization, industry and business, and civil society.

The formalization of civil society into these institutions was accompanied in parallel by two phenomena: on the one hand, the formalization of civil society into environmental non-profits, and on the other, the increasing resistance of other parts of the public sphere – business and politics – to climate action.

In 1989, environmental organizations, led by the Worldwide Fund for Nature (WWF), Environmental Defense Fund (EDF), and Greenpeace International, developed a network in Europe and the United States around climate change known as the Climate Action Network (CAN) (Weart, 2011, p.72). CAN originally started with 63 environmental NGOs from 22 countries (Newell, 2000, p.126) and, as a loosely coordinated network, facilitated exchanges and discussions around climate change action and climate policy, shared strategies on influencing international negotiations, and developed a common platform. They remain to this day an expanded network of 1,300 organizations in over 130 countries (CAN, 2021). Even before the Rio Earth Summit, NGOs within and beyond CAN worked to inform policymakers as well as the general public about the issue. In 1990, Greenpeace published a book entitled *Global Warming: The Greenpeace Report* (Leggett,), which described the science of climate change and its possible consequences. In India at that time, the Centre for Science and Environment was working on its publication *Global Warming in an Unequal World* (Dubash, 2019). At the same time, Friends of the Earth was using its democratic structure to raise public awareness through local groups in countries where it had significant membership (Rahman and Roncerel, 1994, p.245).

This formalization of civil society and its coordinated efforts to raise awareness about the urgency of climate change was accompanied by increased resistance to climate action on the part of business and politics. In their book *Merchants of Doubt* (Oreskes and Conway, 2010) – and the 2014 documentary by the same name – Naomi Oreskes and Erik Conway show how companies with interests in delaying collective action to curb climate change framed the debate as one only of science, technical expertise, and information. These companies used covert tactics, such as funding established scientists, to flood the discourse with claims about scientific disagreement on climate change. By casting doubt on what is in fact a consensual scientific issue – namely that climate change is human-made and requires a global response to reduce greenhouse gas emissions by weaning the economy off fossil fuel-based energy – these 'merchants of doubt' implicitly relied on this distinction between expertise and authority to exclude ordinary citizens from the debate about climate change. As late as 1998, the New York Times published a memo by the American Petroleum Institute that claimed that "uncertainty" about climate change was part of the "conventional wisdom," legitimizing the efforts of powerful companies to sow doubt into the public sphere (Cushman Jr, 1998). Since climate change is not settled, not even by science, the view goes, it would be useless – and perhaps even wrong – for citizens to be politically engaged on the issue. Civil society organizations working on climate, then, would be wrong, ideological, and misguided for pushing for climate action.

It was not until the early 2000s that the tendency for businesses to oppose climate action shifted. It was at that time, in 2002, that the Global Climate Coalition was deactivated, a group bringing together businesses that funded climate-denying so-called science, provided an industry voice opposing climate action, fought the Kyoto Protocol, and opposed action to reduce GHG emissions. Its deactivation was the result of companies leaving the group, with Ford Chrysler leading the exodus in 2000 (Climate Investigations Center, 2019).

It was also in 2002 that the Carbon Disclosure Project (which now goes by "CDP") was launched with the mission to encourage companies to disclose and report environmental data including emissions information for investors (Reid and Toffel, 2009, p.4). Today, after almost 20 years, CDP works with 9,600 companies, 810 cities, and 130 states and regions, disclosing to over 800 investors with more than $110 trillion in AUM and with 147 major buyers with over $4 trillion in purchasing power in global supply chains (CDP, 2019).

Many other organizations have succeeded in mobilizing business, and some companies themselves have taken leadership roles in advancing climate action. This progress occurred in a climate where some of the sectors most foundational to the economy and responsible for most GHG emissions and other destructions of ecosystems and life-systems – in energy, transport, and agriculture – opposed it. This was in a context of increasing efforts to mobilize civil society and strengthening, albeit slowly, political will to tackle the emergency.

Some of the notable successes of civil society organizations over the last three decades can be measured by the achievement of the Paris Agreement in 2015 at the outcome of the 21st Conference of the Parties. This was the first time an agreement was reached between all parties that, with its basis in science, would yield a collective global solution to the climate emergency. The Paris Agreement marked a historic turning point not only for climate action but also for international cooperation and politics, too. In the context of discussing civil society, it is important to note that the path to the Paris Agreement was paved by efforts from civil society as well as from the Parties themselves. Indeed Chapter 3 outlined some of the strides civil society has made in the more recent past: youth movements, climate litigation, direction of action, education, and divestment efforts all mark successes of a movement that started out decades ago with the power of anti-climate action companies and governments against it.

The role of civil society in climate action is multifarious, and no one organization or movement covers all of these possible roles. What is more, most of these roles are aspirational, and when confronted with the concrete conditions of advocacy, campaigning, or direct action, many organizations at times fall short. These roles include:

- Improving the general population's access to climate information and increasing the level of public awareness about climate risk and climate action.
- Acting as "bridges" between research institutions and the general public and decision-makers, translating science in terms that are comprehensible and actionable.
- Giving voice to vulnerable groups. Either through representation or by providing a platform to these groups, CSOs work to ensure acknowledgment of the high vulnerability of these groups in public policy and public awareness, through advocacy processes, other forms of campaigning, or direct action.
- Monitoring progress of various actors in business and politics to ensure that commitments made are delivered.

- Expose scandals and other forms of malpractice to pressure organizations.
- Mainstreaming climate change within developmental, environmental, financial policies.
- Ensuring quality and transparency for diverse actors' participation in various processes of representation or of decision-making.
- Promoting a participatory and inclusive climate movement.
- Actively participating in or facilitating inter-institutional coordination at local, national, or international levels.
- Implementation (or support in implementation) of different projects on the ground.

It is not possible for a single organization to aspire to embody all of these roles; indeed, some of these will be in contradiction with each other. Some organizations will be oriented toward a more consensus-based approach, for instance, by facilitating processes, while others will take a more conflict-based approach, as when they expose greenwashing. Some organizations will serve the interests of a specific constituency; others will work for the benefit of society as a whole (Nilsen, Strømnes and Schmidt, 2017, pp.20–40). Most will tend to agree on the overarching goals of climate action – namely goals deduced from the scientific basis available to all – but there will be divergence on the means to reach them. In short, the climate movement is not homogeneous.

Together, this heterogeneous field has achieved notable successes. However, those successes do not amount to the change we need to see in order to align with requirements based on science, at a minimum, and to transition to a carbon-neutral, inclusive, and fair economy. Rather than analyze all the individual failures, it is more helpful to understand two major structural obstacles faced by the part of civil society that aims to increase climate action. The following two structural features of the climate movement within civil society are presented here as major obstacles to continued impact and the kind of success one might expect, or at least hope, from such a mobilization of resources. Not all organizations face these obstacles in the same ways; there are degrees to which they are subject to them and specific limitations due to the idiosyncrasies of these groups.

Leadership structures: non-diverse and grandfathered

According to a 2018 study, environmental organizations are neither gender nor race diverse: whites comprised more than 80% of the board members of the more than 2,000 groups studied; whites constituted more than 85% of the staff of environmental nonprofits; males occupy about 62% of the board positions while they comprise less than half of the staff of the organizations (Taylor, 2018, p.1). Where women were in positions of leadership, they were very likely to be white. What is more, according to the author of this report, many people in positions of power, statistically more likely to be white and male, have been in their present position for up to forty years, when the movement was nascent. Indeed, in many organizations some of the leadership got into top leadership positions in the late 1970s, early 1980s and have remained in them since, even as the challenges of climate change have evolved and called for equally evolving skills and competences. The author also found that the level of education in environmental or related fields was not a predictor of leadership positions. In her study of the emergence of the conversation movement in the United States in the 19th century, the author also shows how it was led by white urban elites, whose efforts often discriminated against the lower class and were often tied up with slavery and the appropriation of Native lands, the movement benefited from contributions to policymaking, knowledge about the environment, and activism by the poor and working class, people of color, women, and Native Americans without giving them the power commensurate with their input (Taylor, 2016).

Notably, very few organizations report on diversity and inclusion data to this day. 14.5% of the organizations in this study say they engage in some form of diversity, equity, and inclusion activity; 6.8% of the organizations report data on gender diversity; 3.9% of the organizations reveal data on racial diversity; 0.7% of the organizations disclose data on sexual orientation. For a group of organizations that rely on accurate and robust environmental reporting for their mission-driven work, the fact that reporting on their own operations and functioning is so sparse suggests a lack of self-awareness, and more importantly, a lack of genuine engagement with the structural issues that may form obstacles to impact.

Indeed, in an earlier study from 2016, a report further showed that the issue of diversity was not a priority in the field. While most of the almost 300 organizations surveyed expressed a desire to improve on diversity, few had a diversity manager or had formed a diversity committee, and none of the grantmaking foundations studied had a diversity manager (Taylor, 2014, p.4). In the same group, more than two-thirds of the organizations indicated that training programs for minority and low-income residents should be developed, but less than 45% of the organizations would support such training programs.

The point of exploring the structures of leadership is not simply to expose how the field is doing on a metric that has become increasingly popular in a way that may seem to many per se. There are reasons why more diversity and better inclusion are necessary conditions for the success of climate action. Environmental injustice, including climate injustice, has a disproportionate impact on communities of color, on low-income communities, and on the most vulnerable members of society globally. Most evidently, the impacts of climate change take a higher toll on frontline populations – often composed of people of color – than on wealthier people whose lifestyles cause much more environmental damage. Urban neighborhoods predominantly inhabited by black and brown people often lack transportation options, tree canopy, and open spaces that help to mitigate the hardships climate change causes. With its causes in extractivism, it is not surprising that climate change will have disproportionate impacts on the victims of that system too. What is more, however, is that the transition to a decarbonized economy cannot succeed until there are structural changes in society to redress centuries of systemic racism born out of colonialism, slavery, and capitalism. The complex of political and corporate power that is deeply entrenched in the current economy and that evolved over centuries is an obstacle to the changes that are needed to dismantle the fossil fuel interests responsible for the climate crisis (Holthaus, 2020; McKibben, 2020).

Where gender diversity is concerned, there is much evidence showing that women are not only among the most touched but also among the most active and most effective agents of change for climate solutions. According to the UN, women around the world are more likely to die when natural disasters such as heatwaves, droughts, rising sea levels, and extreme storms strike than men and are more likely to suffer adverse effects in the aftermath, too (Peihong et al., 2016, p.IX). To illustrate, the study found that women in China will be worse off than men in the country in the face of environmental change: women comprise 70% of the agricultural workforce in China and have less access to income, land, technology, loans, and employment options outside of farm work than their male peers. Beyond a lack of access to resources, about 80% of women in the country were unfamiliar with disaster emergency plan. These factors explain why women are particularly vulnerable to the impacts of climate change.

On the other hand, women have been shown to be powerful agents of change in the fight against climate change. Consider that, according to the UN Food and Agriculture Organization, if women farmers were to have equal access to the same resources, such as trainings, financing, and property rights, as their male counterparts, their crop yields could increase by 20–30%, strengthening their livelihoods and the welfare of their households (FAO, 2011, p.vi). When women are able to earn their own wages, they typically reinvest 90% of those earnings back into their families and communities (FAO, 2016).

What does this data have to do with diversity in the climate movement itself? First of all, lack of representation is likely to result in lack of adequate consideration for these particular stakeholders. In addition, failing to recognize gender-based inequity and injustice in the movement itself may very well predict that this injustice will not be addressed with adequate resources and at an adequate scale in the solutions sought.

Inconsistency and inauthenticity

Organizations sometimes present surprising portfolios of solutions, either because they contain clear tensions or because their consequences lead to contradictions with their primary missions. At its inception in the 1970s, the EDF created a detailed energy forecast, according to which California would reduce electricity consumption rather than need to consider building more capacity with nuclear for instance (Roe, 1984). When it turned out that California could not reduce consumption, the state turned to natural gas (Shellenberger, 2018).

In the following decade, EDF pushed for the deregulation of electricity markets to the benefit of natural gas companies. At the same time, NRDC also advocated deregulation and even publicly supported Enron. NRDC's energy leader even claimed, that "on environmental stewardship, our experience is that you can trust Enron" (Beder, 2003, p.6). Meanwhile, Enron executives were notoriously defrauding investors of billions.

Decades later, in 2009–2011, EDF and NRDC associated with lobbyists and lawyers that pushed for cap-and-trade climate legislation that would have created a carbon-trading market worth upwards of $1 trillion (for comparison, the total revenue of the US electricity market in 2018 was $400 billion) (EIA, 2019), which can be said to amount to a carbon tax structured so that it is captured by private interests. A different scheme could structure this tax so as to be reinvested in the transition to a new economy. Notably, Nobel Prize winner Al Gore, co-founded a company called Generation Investment Management with former Goldman Sachs leadership to invest in carbon offsets and do just that – he is currently Chairman (Generation, 2017).

Today, EDF works with the world's largest multinational oil and gas companies, and the demands they are making in the field of regulation would benefit highly capitalized companies such as these (Ratner, 2019). Along with the Sierra Club and NRDC, organizations similar in mission, stature, and influence as EDF all take in more than half a billion dollars from donors that include billionaire coal and natural gas investors such as Tom Steyer and Michael Bloomberg (EDF, 2019; NRDC, 2019, p.20; Sierra Club Foundation, 2019). These two supporters of climate action have become more prominent in the field and as funders of CSOs in more recent years in the United States since responsibility has been relinquished by most politicians in the country. They are both billionaires and have both made their fortunes by investing in fossil fuels – indeed it is only as Tom Steyer prepared for his presidential run in 2020 that he announced he would divest. Bloomberg made no such announcement when he decided to run on the Democratic ticket that year, in contrast. The contradictions between the financial interests of both of these individuals, their short-lived but nonetheless disproportionately funded presidential campaigns, and their philanthropy for climate action has, at the very least, caused some to distrust their intentions and the likelihood that the organizations they support are adequately independent of these funders. While Bloomberg has been critical in closures of coal plants in the United States (Grunwald, 2015), his opposition to banning fracking (Irfan, 2020), his anti-union stances (Gardiner, 2019), his resistance to a vision of climate action that integrates social and economic justice such as the Green New Deal (DiChristopher, 2019), and his own unstraightforward stance on natural gas (Aton, 2020) cast doubt on how the organizations he supports may be steered or worse, have their hands tied by his donations.

This is one example of a history of advocacy and associations that appear inconsistent with the mission of the climate movement, namely, following the requisites of science to transition the world economy away from fossil fuels toward a net-zero economy before the window of opportunity closes.

Other organizations have been the subjects of scandals that demonstrate other forms of inconsistency. Greenpeace, for instance, was revealed in 2014 to have engaged in financial mismanagement and to have vetted commuting by air travel for some of its leadership (Vaughan, 2014). In 2018, the UN's environment chief, Erik Solheim, resigned after it came to light that he had spent half a million US dollars in 22 months of tenure on air travel and hotels (Carrington, 2018). Indeed, the UN conference COP drew an average of 16,500 participants between 2009 and 2016 (latest figures available) (UNFCC, 2018) in places as disparate as Lima (Peru), Warsaw (Poland), Durban (South Africa), or Doha (Qatar). These participants overwhelmingly travel by air. Meanwhile, all of these organizations, including the hundreds of organizations who attend COP, all claim that air travel is one of the most egregious contributors to rising GHG emissions, since it accounts for about 2.5% of global CO_2 emissions annually – and could account for about a quarter of the world's carbon budget by 2050 according to projections (Pidcock and Yeo, 2016).

There are many paths to reaching the goal of reaching a just and net-zero economy and complementary ones at that. However, some will create further obstructions, and it stands to reason that furthering the interests of fossil fuel companies and other investors who will not reinvest in this transition seems like such an obstruction. It further stands to reason that if the staunchest advocates for the transformation of the economy cannot find ways to alter their practices when these are known to be extremely deleterious to the mission, the path to transformation is obstructed. In addition to creating distrust and skepticism about the climate movement, which has known, as it is, much well-financed opposition, it also raises questions as to whether the current organization of civil society organizations working in this field is fit for purpose.

In the face of these limitations and indeed in the face of the insufficient – from the point of view of what science calls for – impact of CSOs on climate, we can look to different forms of organizing and action for more hopeful avenues. Youth movements are not new, but they have gained momentum and influence in recent years. In particular, they show signs of being able to avoid the two structural obstacles of grandfathering and lack of diversity on the one hand and of inauthenticity and inconsistency on the other. Youth activists today are different from their predecessors since they are louder and more coordinated, mostly thanks to their use of social media and the feedback look created by this increased visibility (Fisher, 2019, pp.430–431).

Marking a shift from older generations, young persons are more likely to take the issue of climate and of climate science seriously. Six out of seven youths aged 13 to 17 surveyed in a 2019 poll believed that "human activity is causing climate change" and four out of five called it a "crisis" (The Washington Post, 2019, p.6). Among high school students, 40% say they have taken action to reduce their own carbon footprint, while 80% say they have participated in direct action.

In particular, the youth movement of today has managed to mobilize other parts of the population by making the argument that climate action today is an obligation to generations of tomorrow. A 2016 survey of 1,860 people in the United Kingdom found that 61% were willing to pay up to £20 ($25) a month to prevent climate change–related deaths in 2050, 2080, and 2115 (Graham et al., 2019, p.113). And participants in a 2017 study conducted in Lisbon and Adelaide, Australia, were willing to spend just as much money to prevent negative climate impacts on future generations as they were to protect themselves (Everuss et al., 2017, p.1). No doubt the fact that youth activists are fighting for their own future – and against the inaction of those in power – are reasons to act and mobilize are difficult to ignore for adults.

Youth movements avoid the pitfall of inauthenticity and grandfathering. The latter for obvious reasons of age, though of course privilege and discrimination are at play in this forum as much as in any other (Burton, 2019). But they do not respond to the same constraints and interests as their adult counterparts, who must protect their careers or protect their funding sources. Youth activists approach the simplest of motives: fighting for their very own future and well-being, and they can do so without the filters and other burdens of vested interests. They are able to be authentic in ways that more formal parts of civil society cannot.

It is easy to overstate the impact of youth activists given their formidable visibility. However, beyond mass mobilization, which is no small feat, youth activists can be credited with much significant action on the ground. In the United States, the Sunrise Movement, a youth-led movement launched in 2017 with the purpose of supporting the elections of climate-progressive candidates in state and local elections. About half of the candidates they endorsed won their seats in 2018 midterm elections, including Green New Deal architect and champion Alexandria Ocasio-Cortez (Medium, 2018). Another notable success is their campaign for including a debate focused on climate in the primary campaign of the Democratic Party (Brady, 2019). In Africa, where the youth-led climate marches gathered hundreds of thousands of citizens in 2019, young women have also distinguished themselves by leading change at different levels – either by promoting the educations of local communities on the links between the impacts of climate change and their socio-economic prospects, organizing plastic pick-ups, and representing the interests of the regions on the international scene (Hanson, 2019). In Latin America, teenagers have also demonstrated leadership and commitment in the fight against climate change, for instance, by launching the "1,000 actions for a change" in 2019 and mobilizing youth in more than 20 countries in the region (UNICEF, 2020). Groups of youth all over the world have sued their governments for climate inaction, and in Pakistan, Colombia, and South Africa, they have taken their governments to court and won (Parker, 2019).

As is to be expected from grassroots movements, it is not yet clear what the aggregate impact of these initiatives will be, but it is clear that this scale of mobilization is unprecedented. What is more, over 40% of the world population is under the age of 25, and this category is growing especially in regions that are hardest hit by the impacts of climate change, South Asia and Sub-Saharan Africa (Khokhar, 2017). Not only are youth activists a formidable influence today, but they are also the government, corporate, and institutional leaders of tomorrow and are likely to realign power in unprecedented ways.

Civil society, climate action, and business leadership

What does this mean for business? As we noted, civil society is heterogeneous, and hence there is no one single rule for engaging with civil society on climate change. However, given the structural features of civil society we have presented, business must ensure the following before it engages with CSOs:

- **Due diligence on organizations** that a company engages with, with questions such as:
  - Does this organization take funding and support from fossil fuel companies?
  - What is the organization's record on key areas of climate action?
  - What is the organization's leadership history?
  - What is the diversity of the organization's staff, and how is this reflected in the actions of the organization?
  - Does the organization abide by the kinds of recommendations it gives to business?
  - How long has the organization been "in the business" of climate change, and what has it achieved?
  - Does the organization reflect the shift in values that can be observed in society, including a shift toward decentralized power or prioritizing diversity and inclusion in its own structure and in its campaigns and initiatives?
  - Does the organization work in other areas, and what is its record in those?

- **Due diligence on the company's** own needs and motivations, with questions such as:
  - Does the company need to communicate about current actions to specific stakeholders?
  - Does the company need to assess its performance with a view to improvement?
  - Does the company need to launch an initiative?
  - Does the company need to engage its leadership?
  - Is the company willing to keep to requisite distance from civil society in order to be criticized publicly for falling short of civil society's expectations?
  - Is the company prepared to consider areas of operations and structural issues that are not typically considered to be the remit of climate action such as its own record on inclusion and diversity?
  - Is the company seeking exposure for current initiatives only? Or is the company seeking transformative change with the support of external stakeholders such as CSOs?

- Is engaging with a CSO the most effective way to achieve the impact the company seeks?
- Where does the company's power lie, and how can it be activated in support of this given cause or mission?

These questions do not exhaust the possibilities of performing a robust due diligence. They show that two overarching principles govern these rules of engagement: impact and authenticity, and these will interplay with varying degrees of priority depending on context. Through this kind of due diligence, companies can both shield themselves from the structural weaknesses of civil society itself, avoid reputational risk, and ensure that they reach their goals. This kind of due diligence can help avoid these pitfalls by:

Confusing power and powerful gestures

The distinction between power and powerful gestures is important here. Because of the relations they hold with their stakeholders, companies hold genuine power over large swathes of the world population: employees through wages and time management, suppliers through contracts, local communities via their use of local resources, governments through lobbies, consumers through their offer and pricing, and so on and so forth. Genuine change is the kind of change that transforms these power relations rather than gesturing toward this change. This is sometimes called "greenwashing" in the context of environmental action, but this confusion is not limited to that field. Consider companies that declared, in the wake of Black Lives Matter uprisings in the United States, their solidarity with the movement. Upon inspection, some of these companies turned out not to have invested resources in diversity and inclusion, to have ties with stakeholders that actively oppose the movement's demands, or to display endemic discrimination in their activities. There are parallels between these actions and how climate is approached by companies, who often see climate as an issue to be vocal about rather than to be active about, especially as the issue becomes more and more important to stakeholders. See tables below for examples of this obfuscation in the context of BLM and in the context of climate action.

Examples of powerful gestures vs. power confusion in the BLM context

- Amazon tweeted its support for BLM on May 31, 2020, while at the same time failing to address working conditions that make Amazon warehouse and delivery workers some of the most at-risk and precarious workforces in many of the countries where it operates. At the time of the tweet, Amazon was also cooperating with police departments on the development and sale of facial recognition technology that is known to embed racial discrimination and which would exacerbate the racial biases in police activities, too (Ng, 2019). Amazon is the second largest employer of black people in the United States, who make up 27% of its workforce (Amazon, 2021b).

- On May 29, Target CEO expressed his solidarity for the BLM movement, failing to bring up that Target had cooperated with police departments across the United States, including in Minneapolis, where George Floyd's catalytic murder took place. This cooperation has included $300,000 to the Minneapolis Police Department to set up surveillance cameras across a 40-block radius in the city (Giles, 2012). In 2011, Target added a forensics crime lab to their campus in Minnesota to collect high-resolution images from their surveillance cameras, which were channeled to local police at no cost (Target, 2012). What is now called their "Public Safety Focused Grant" was once termed "Law Enforcement Grant Program." Such cooperation with police departments across the United States is inconsistent with support for the BLM movement, which, among other things, calls for a radical transformation of the police up to and including defunding it. Of course, a statement by a CEO that reaches millions is a powerful gesture, but it is not where the real power of a company lies.
- Uber issued a statement outlining steps to support black businesses and to introduce accountability at the executive level on inclusion (Uber, 2020). Lyft also expressed support to the BLM movement and condemned systemic racism in a statement from the CEO, who also pledged $500,000 donations in rides to civil rights activists (Green and Zimmer, 2020). This is about 0.1% of the company's annual revenue. DoorDash pledged $500,000 to Black Lives Matter (Xu, 2020). The three companies together spent over $90 million in 2019 on lobbying against the right of workers to join unions, earn minimum wage, or collect healthcare benefits (Conger, 2019). What is more, the algorithms of the two rideshare companies result in higher charges for rides to non-white neighborhoods, embedding social and racial discrimination in the very heart of the business (Pandey and Çalişkan, 2020, p.1).
- Walmart has pledged $100 million to address systemic racism (Walmart, 2020). Walmart is the largest employer of black people in the United States; they make about 21% of its workforce and are highly concentrated in its retail operations (Walmart, 2018). Walmart has been staunchly anti-union for decades and is known to pay retail employees poverty wages (Bomey, 2018).

Examples of powerful gestures vs. power confusion in the climate change context

- In September 2019, Jeff Bezos unveiled a new corporate "Climate Pledge" and signed up Amazon as the first participant (Amazon, 2019). With this pledge, Amazon set a deadline to reach 100% renewable energy – 2030 – and net zero by 2040. The company did not, importantly, provide more detail about its own current impacts, making these targets impossible to assess. It is also noteworthy that Amazon's renewable commitment covers its own operations and electricity use only and excludes its supply chain – which accounts

for more than 75% of its carbon footprint (Amazon, 2020). In addition, Amazon sells machine learning and other AI technologies to oil companies like BP, Shell (Amazon, 2021a).
- Microsoft sells services to fossil fuel companies that help them drill for more oil and gas – as much as 50,000 barrels more a day for Exxon in Texas' Permian Basin (Microsoft, 2019a). In 2019, Microsoft launched an AI Center for Excellence to assist the oil industry's digital transformation (Microsoft, 2019b) and sponsored an oil and gas conference the exact same week that it announced its new climate commitment (Stone, 2020).
- BlackRock CEO publicly and vocally stated that climate change could lead to a "fundamental reshaping of finance" and that his company would place sustainability at the heart of its investment strategy. In May 2020, the world's largest asset manager refused to back landmark environmental resolutions at two big Australian oil companies that would have called for Woodside Energy and Santos to set targets in line with the Paris Agreement. The resolutions had, respectively, above 50% and 43% support.

Where climate action is concerned, eschewing this confusion – or even obfuscation – will involve prioritizing concrete action versus statements and ensuring consistency within these action portfolios. While there is power in communication and making ambitious climate action the norm through considered dissemination, action must precede and fully support public claims. This is related to the pitfall of greenwashing, which consists in conveying a false impression or even providing misleading information about a company's environmental performance. Greenwashing often includes the notion that companies are intentionally deceitful about these issues, while what we are talking about here is the kind of inconsistency that can arise not only out of deceit but out of negligence too. The pitfall we are describing here includes greenwashing but is not limited to it.

Reciprocally co-opting civil society

In order for civil society organizations to fulfill their variety of roles – including exposing companies, holding them accountable, and guiding them in the way to having more impact – a certain independence with companies must hold. We saw earlier in this chapter how some large environmental NGOs receive funding from companies in high-emitting industries and often end up engaging in advocacy for solutions that are amenable to these sectors at the cost of being the right solutions for climate. NGOs with a business focus are many and they are not all problematic. They become problematic when this proximity to NGOs stands in the way of NGOs acting to fulfill their roles.

One development that has emerged from this close cooperation where NGOs opt to work with corporations and vice-versa is that there has been much effort into market-friendly solutions and consumer-driven activism such as certification and labeling. Rather than challenging business, as usual, this kind of relationship serves to legitimize it and to favor incremental change. Companies become allies rather than adversaries and become seen as change agents that make the economy allegedly more climate-compatible without challenging the deep structures that stand in the way. Forgoing the ambition of radical and structural change, NGOs that are close to companies in this way take on a different role that may be useful in some limited respects but that will not yield the change and leadership that is required by the climate emergency.

NGOs that work in this vein have many reasons to do so. Companies are sources of funding and for organizations with employees on a payroll and operating costs, these sources are precious. Furthermore, the reach and visibility companies afford these organizations can be very valuable to raise awareness about their activities and mission within the general public. However, receiving funding from companies can place organizations in a position where they must prioritize the needs of the funder ahead of the needs of the mission. If companies recognize that the primary mission with regard to climate change is to avert its worst impacts by transforming the economy, it is unlikely that these organizations will continue to be seen as the agents of change they are sometimes perceived to be.

There is no reason, conceptually, that companies cannot be allies in the transformation of the economy we need. However, our argument has been that this transformation requires paradigmatic and structural changes and not incremental changes. Companies can be allies, then, only if they engage in activities that will promote these kinds of changes. It is unlikely that the parts of civil society that promote business-friendly "business as usual with small tweaks" approaches will in fact harness the potential that companies become powerful allies. There is no reason, either, why these types of organizations do not have a purpose in civil society. One can see this kind of "co-opted" organization as a helpful step for companies to gain awareness and information about climate action and the need for leadership. Importantly, however, this kind of organization – one that has lost the requisite critical distance from what they are supposed to challenge and change – is but one of many pushes that the system needs in order to shift to a new climate economy.

Conclusion: allies or adversaries?

As the role and power of civil society changes, so does the business and its relationship to it. This book has argued that business has the potential to lead the radical transformation of the economy in the face of the climate emergency only if it transforms itself. In order to do so, it must avoid power dynamics that will cement the status quo. Parts of civil society run the risk of doing just that.

The adversarial model of civil society is helpful insofar as it keeps checks on business. It is important to maintain the distance necessary for critical engagement while recognizing that constructive relationships with civil society organizations are possible and necessary for very specific functions. In a period where stakeholder trust is a valuable asset but also one that is at risk, protecting conditions of transparency and trustworthiness is a company's most important tactic. The responsibility of navigating the space of civil society befalls companies. Leadership will require them to recognize and maintain the appropriate distance with CSOs, protecting their independence and creating conditions of trust. It is also their responsibility to respond to the claims of CSOs not with gestures but with actions.

Note

1 Many other women were leaders in the Civil Rights movement and have often been overlooked by the history books: Ella Baker, a labor organizer and longtime leader in the Southern Christian Leadership Conference; Septima Poinsette Clark, often called the "queen mother" of civil rights, educator and National Association for the Advancement of Colored People activist; Fannie Lou Hamer, co-founder of the Mississippi Freedom Democratic Party; Vivian Malone Jones who defied segregationist Alabama Gov. George C. Wallace to enroll in the University of Alabama in 1963 and later worked in the civil rights division of the US Justice Department. These women are just a few of the black leaders that fought for more equality in the United States, and countless Black women whose name is not recorded were also essential to the advancement of this cause.

References

AFDB (African Development Bank Group) (2012) *Framework for Enhanced Engagement with Civil Society Organizations*. Available at: https://www.afdb.org/sites/default/files/documents/policy-documents/framework_for_enhanced_engagement_with_civil_society_organizations1_0.pdf (Accessed: 20 March 2021)

Amazon (2019) *Amazon Co-Founds The Climate Pledge, Setting Goal to Meet the Paris Agreement 10 Years Early* [Press release]. 19 September. Available at: https://press.aboutamazon.com/news-releases/news-release-details/amazon-co-founds-climate-pledge-setting-goal-meet-paris (Accessed: 20 March 2021)

Amazon (2020) *Sustainable Operations Carbon Footprint* [Online]. Amazon Sustainability. Available at: https://sustainability.aboutamazon.com/environment/sustainable-operations/carbon-footprint (Accessed: 20 March 2021)

Amazon (2021a) AWS Energy [Online]. AWS Amazon. Available at: https://aws.amazon.com/energy/?energy-blog.sort-by=item.additionalFields.createdDate&energy-blog.sort-order=desc (Accessed: 20 March 2021)

Amazon (2021b) Our Workforce Data [Online]. Amazon. Available at: https://www.aboutamazon.com/news/workplace/our-workforce-data (Accessed: 20 March 2021)

Aton, A. (2020) 'Bloomberg Confronts Pro-Gas Efforts of Past in Building Plan,' *E&E News*, 16 January [Online]. Available at: https://www.eenews.net/stories/1062091595 (Accessed: 20 March 2021)

Baker, J. and Haug, H. (2002) *The Kosovo Women's Initiative An Independent Evaluation*. Available at: https://www.unhcr.org/afr/3db019784.pdf (Accessed: 20 March 2021)

Beder, S. (2003) 'How Environmentalists Sold Out to Help Enron,' *PR Watch*, 10(3) [Online]. Available at: http://www.herinst.org/sbeder/privatisation/environ.pdf (Accessed: 20 March 2021)

Bomey, N. (2018) 'Walmart Boosts Minimum Wage again, Hands Out $1,000 Bonuses,' *USA Today Money*, 11 January [Online]. Available at: https://eu.usatoday.com/story/money/2018/01/11/walmart-boosts-minimum-wage-11-hands-out-bonuses-up-1-000-hourly-workers/1023606001/ (Accessed: 20 March 2021)

Brady, J. (2019) 'Activists Push Democrats on Climate Change, a New Priority for Party's Base,' *NPR*, 22 August [Online]. Available at: https://www.npr.org/2019/08/22/753122273/activists-push-democrats-on-climate-change-a-new-priority-for-partys-base?t=1616515603375 (Accessed: 20 March 2021)

Burton, N. (2019) 'Meet the Young Activists of Color Who Are Leading the Charge against Climate Disaster,' *Vox*, 11 October [Online]. Available at: https://www.vox.com/identities/2019/10/11/20904791/young-climate-activists-of-color (Accessed: 20 March 2021)

CAN, Climate Action Network International (2021) About Can [Online]. Climate Action Network International. Available at: https://climatenetwork.org/overview/ (Accessed: 20 March 2021)

Carrington, D. (2018) 'UN Environment Chief Resigns after Frequent Flying Revelations,' *The Guardian*, 20 November [Online]. Available at: https://www.theguardian.com/environment/2018/nov/20/un-environment-chief-erik-solheim-resigns-flying-revelations (Accessed: 20 March 2021)

CDP (2019) What We Do [Online]. CDP. Available at: https://www.cdp.net/en/info/about-us/what-we-do (Accessed: 20 March 2021)

Climate Investigations Center (2019) 'Global Climate Coalition: Climate Denial Legacy Follows Corporations,' *Climate Investigations*, 25 April [Online]. Available at: https://climateinvestigations.org/global-climate-coalition-industry-climate-denial/ (Accessed: 20 March 2021)

Conger, K. (2019) 'Uber, Lyft and DoorDash Pledge $90 Million to Fight Driver Legislation in California,' *The New York Times*, 29 August [Online]. Available at: https://www.nytimes.com/2019/08/29/technology/uber-lyft-ballot-initiative.html (Accessed: 20 March 2021)

Cushman Jr, J.H. (1998) 'Industrial Group Plans to Battle Climate Treaty,' *The New York Times*, 26 April [Online]. Available at: https://www.nytimes.com/1998/04/26/us/industrial-group-plans-to-battle-climate-treaty.html (Accessed: 20 March 2021)

DiChristopher, T. (2019) 'Mike Bloomberg Says Alexandria Ocasio-Cortez's Green New Deal "Stands No Chance" in the Senate – so He's Offering an Alternative,' *CNBC*, 5 March [Online]. Available at: https://www.cnbc.com/2019/03/05/mike-bloomberg-says-green-new-deal-stands-no-chance.html (Accessed: 20 March 2021)

Dubash, N.K. (2019) *India in a Warming World: Integrating Climate Change and Development* E-book format [online]. Available at https://oxford.universitypressscholarship.com/view/10.1093/oso/9780199498734.001.0001/oso-9780199498734-chapter-5 (Accessed: 20 March 2021)

EDF (2019) *New Energy New solutions Annual Report 2019*. Available at: https://www.edf.org/sites/default/files/content/2019_EDF_Annual_Report.pdf (Accessed: 20 March 2021)

EIA (2019) *2019 Total Electric Industry- Revenue (Thousands Dollars)* [Online]. EIA. Available at: https://www.eia.gov/electricity/sales_revenue_price/pdf/table3.pdf (Accessed: 20 March 2021)

European Union (2010) Glossary of Summaries. Civil Society Organisation. EUR-Lex. Available at: https://eur-lex.europa.eu/summary/glossary/civil_society_organisation.html (Accessed: 20 March 2021)

Everuss, L. et al. (2017) 'Assessing the Public Willingness to Contribute Income to Mitigate the Effects of Climate Change: A Comparison of Adelaide and Lisbon,' *Journal of Sociology*, 53(1) [online]. Available at: https://www.researchgate.net/publication/313727060_Assessing_the_public_willingness_to_contribute_income_to_mitigate_the_effects_of_climate_change_A_comparison_of_Adelaide_and_Lisbon (Accessed: 20 March 2021)

FAO (2011) *The State of Food and Agriculture Women in Agriculture: Closing the Gender Gap for Development*. Available at: http://www.fao.org/3/i2050e/i2050e.pdf (Accessed: 20 March 2021)

FAO (2016) *Women Hold the Key to Building a World Free from Hunger and Poverty* [Press release]. 16 December. Available at: http://www.fao.org/news/story/en/item/460267/icode/ (Accessed: 20 March 2021)

Fisher, D.R. (2019) 'The Broader Importance of #FridaysForFuture,' *Nature Climate Change*, 9 [online]. Available at: https://www.nature.com/articles/s41558-019-0484-y (Accessed: 15 November 2020)

Fulton, S. (2020) 'Black Lives Matter Founders Alicia Garza, Patrisse Cullors and Opal Tometi,' *TIME*, 22 September [Online]. Available at: https://time.com/collection/100-most-influential-people-2020/5888228/black-lives-matter-founders/ (Accessed: 20 March 2021)

Gardiner, A. (2019) 'A 'Fixer' or a 'Bully': New Yorkers Have Opinions on Bloomberg as Mayor,' *The New York Times*, 12 November [Online]. Available at: https://www.nytimes.com/2019/11/12/reader-center/bloomberg-new-york-city.html (Accessed: 20 March 2021)

Gayle, D. (2018) 'Timeline: Oxfam Sexual Exploitation Scandal in Haiti,' *The Guardian*, 15 June [Online]. Available at: https://www.theguardian.com/world/2018/jun/15/timeline-oxfam-sexual-exploitation-scandal-in-haiti (Accessed: 20 March 2021)

Generation (2017) Generation Our People [Online]. Generation. Available at: https://www.generationim.com/firm-overview/our-people/ (Accessed: 20 March 2021)

Giles, B. (2012) 'Minneapolis' SafeZone,' *Security Info Watch*, 18 April [Online]. Available at: https://www.securityinfowatch.com/home/article/10702225/minneapolis-publicprivate-surveillance-effort-with-target-corp (Accessed: 20 March 2021)

Graham, H. et al. (2019) 'Willingness to Pay for Policies to Reduce Future Deaths from Climate Change: Evidence from a British Survey,' *Public Health*, 174 [online]. Available at: https://www.sciencedirect.com/science/article/pii/S003335061930191X (Accessed: 20 March 2021)

Green, L.D. and Zimmer, J. (2020) A Note from Our Co-Founders: A Call to Action for Each of Us [Blog] *Lyft Blog*. Available at: https://www.lyft.com/blog/posts/a-note-from-our-co-founders-a-call-to-action-for-each-of-us (Accessed: 21 August 2020)

Griffin, P. (2019) '#MeToo, White Feminism and Taking Everyday Politics Seriously in the Global Political Economy', *Australian Journal of Political Science*, 54(4) [online]. Available at: https://www.tandfonline.com/doi/abs/10.1080/10361146.2019.1663399?src=recsys&-journalCode=cajp20 (Accessed: 20 March 2021)

Grunwald, M. (2015) 'Inside the War on Coal,' *Politico*, 26 May [Online]. Available at: https://www.politico.com/agenda/story/2015/05/inside-war-on-coal-000002/ (Accessed: 20 March 2021)

Hanson, J. (2019) '3 Young Black Climate Activists in Africa Trying to Save the World,' *Greenpeace*, 28 October [Online]. Available at: https://www.greenpeace.org.uk/news/black-history-month-young-climate-activists-in-africa/ (Accessed: 20 March 2021)

Holthaus, E. (2020) 'The Climate Crisis Is Racist. The Answer Is Anti-Racism,' *The Correspondent*, 28 May [Online]. Available at: https://thecorrespondent.com/496/the-climate-crisis-is-racist-the-answer-is-anti-racism/65663772944-cc6110b0 (Accessed: 20 March 2021)

Irfan, U. (2020) 'The Best Case for and against a Fracking Ban,' *Vox*, 7 October [Online]. Available at: https://www.vox.com/energy-and-environment/2019/9/12/20857196/kamala-fracking-ban-biden-climate-change (Accessed: 20 March 2021)

Khokhar, T. (2017) Chart: How Is the World's Youth Population Changing? [Blog] *World Bank Blogs*. Available at: https://blogs.worldbank.org/opendata/chart-how-worlds-youth-population-changing (Accessed: 21 August 2017)

Koch, D. (2009) *Aid from International NGOS: Blind Spots on the Aid Allocation Map*. London; New York: Routledge.

Kroc Institute (2019) *Hacia la paz sostenible por el camino de la igualdad de género*. Available at: https://keough.nd.edu/wp-content/uploads/2019/12/120519_informe_genero_digital.pdf (Accessed: 20 March 2021)

Lamb, J. (2020) 'Six Teens React to Massive Protest They Organized,' *News Channel 5*, 5 June [Online]. Available at: https://www.newschannel5.com/news/six-teens-react-to-massive-protest-they-organized (Accessed: 20 March 2021)

Larbi Sadiki (eds.) (2017) *Routledge Handbook of the Arab Spring: Rethinking Democratization*. London: Routledge.

Leggett, J.K. (1990) *Global Warming: The Greenpeace Report*. Available at: https://journals.sagepub.com/doi/10.1177/027046769201200290?icid=int.sj-abstract.similar-articles.1 (Accessed: 20 March 2021)

McKibben, B. (2020) 'Racism, Police Violence, and the Climate Are Not Separate Issues,' *The New Yorker*, 4 June [Online]. Available at: https://www.newyorker.com/news/annals-of-a-warming-planet/racism-police-violence-and-the-climate-are-not-separate-issues (Accessed: 20 March 2021)

Medium (2018) 'Sunrise Movement Announces First Round of Endorsed Candidates,' *Medium*, 8 August [Online]. Available at: https://medium.com/sunrisemvmt/sunrise-movement-announces-first-round-of-endorsed-candidates-e3638d1a5e9b (Accessed: 20 March 2021)

MeToo Movement (2021) Statistics [Online]. MeToo. Available at: https://metoomvmt.org/learn-more/statistics/ (Accessed: 20 March 2021)

Microsoft (2019a) 'ExxonMobil to Increase Permian Profitability through Digital Partnership with Microsoft,' *Microsoft News Center*, 22 February [Online]. Available at: https://news.microsoft.com/2019/02/22/exxonmobil-to-increase-permian-profitability-through-digital-partnership-with-microsoft/ (Accessed: 20 March 2021)

Microsoft (2019b) 'Microsoft Announces 'AI Centre of Excellence' at ADIPEC 2019, to Accelerate Innovation across Energy Sector,' *Microsoft News Center*, 12 November [Online]. Available at: https://news.microsoft.com/en-xm/2019/11/12/microsoft-announces-ai-centre-of-excellence-at-adipec-2019-to-accelerate-innovation-across-energy-sector/ (Accessed: 20 March 2021)

Millman, O. (2020) 'Amazon Threatened to Fire Employees for Speaking out on Climate, Workers Say,' *The Guardian*, 2 January [Online]. Available at: https://www.theguardian.com/technology/2020/jan/02/amazon-threatened-fire-employees-speaking-out-climate-change-workers-say (Accessed: 20 March 2021)

NRDC (2019) *In Fifty Years 2019 Annual Report*. Available at: https://www.nrdc.org/finances-and-annual-report (Accessed: 20 March 2021)

Newell, P. (2000) *Climate for Change: Non-State Actors and the Global Politics of the Greenhouse Effect*. Cambridge: Cambridge University Press.

Nilsen, H. R., Strømnes, K. and Schmidt, U. (2017) 'A Broad Alliance of Civil Society Organizations on Climate Change Mitigation: Political Strength or Legitimizing Support?,' *Journal of Civil Society*, 14(1) [Online]. Available at: https://www.tandfonline.com/doi/full/10.1080/17448689.2017.1399596 (Accessed: 15 November 2020)

Nulman, E. (2015) *Climate Change and Social Movements Civil Society and the Development of National Climate Change Policy*. London, UK: Palgrave Macmillan.

Oreskes, N. and Conway, E.M. (2010) *Merchants of Doubt: How a Handful of Scientists Obscured the Truth on Issues from Tobacco Smoke to Global Warming*. London: Bloomsbury.

Pandey, A. and Çalişkan, A. (2020) 'Iterative Effect-Size Bias in Ridehailing: Measuring Social Bias in Dynamic Pricing of 100 Million Rides,' *ArXiv* [online]. Available at: https://arxiv.org/pdf/2006.04599.pdf (Accessed: 15 November 2020)

Parker, L. (2019) 'Kids Suing Governments about Climate: It's a Global Trend,' *National Geographic*, 26 June [Online]. Available at: https://www.nationalgeographic.com/environment/article/kids-suing-governments-about-climate-growing-trend (Accessed: 20 March 2021)

Peihong, Y. et al. (2016) *Gender Dimensions of Vulnerability to Climate Change in China*. Available at: https://www2.unwomen.org/-/media/field%20office%20eseasia/docs/publications/2016/12/deliverable%207-english.pdf?v=1&d=20161208T095438 (Accessed: 20 March 2021)

Phipps, A. (2019) '"Every Woman Knows a Weinstein": Political Whiteness and White Woundedness in #MeToo and Public Feminisms around Sexual Violence', *John Hopkins University Press*, 31(2) [online]. Available at: https://muse.jhu.edu/article/736793/pdf (Accessed: 20 March 2021)

Pidcock, R. and Yeo, S. (2016) 'Analysis: Aviation Could Consume a Quarter of 1.5C Carbon Budget by 2050,' *CarbonBrief*, August 8 [Online]. Available at: https://www.carbonbrief.org/aviation-consume-quarter-carbon-budget (Accessed: 20 March 2021)

Rahman, A. and Roncerel, A. (1994) 'A View from the Ground Up', in I.M. Mintzer and J.A. Leonard (eds.) *Negotiating Climate Change: The Inside Story of the Rio Convention*, Cambridge: Cambridge University Press, pp. 239–273.

Ratner, B. (2019) EDF and ExxonMobil Discuss Technology and Regulation to Reduce Methane Emissions [Blog] *Environmental Defense Fund Blog*. Available at: http://blogs.edf.org/energyexchange/2019/03/13/edf-and-exxonmobil-discuss-technology-and-regulation-to-reduce-methane-emissions/ (Accessed: 21 August 2020)

Reid, M.E. and Toffel, M.W. (2009) *Responding to Public and Private Politics: Corporate Disclosure of Climate Change Strategies*. Harvard Business School Paper No 09-019. Available at: https://www.hbs.edu/ris/Publication%20Files/09-019_099ceb6c-44e0-4471-b712-9182fde7b806.pdf (Accessed: 20 March 2021)

Roe, D. (1984) *Dynamos and Virgins*. New York, NY: Random House.

Shellenberger, M. (2018) 'Jerry Brown's Secret War on Clean Energy,' *Environmental Progress*, 11 January [Online]. Available at: https://environmentalprogress.org/big-news/2018/1/11/jerry-browns-secret-war-on-clean-energy (Accessed: 20 March 2021)

Sierra Club Foundation (2019) *Powering Our Future Annual Report 2019*. Available at: https://vault.sierraclub.org/foundation-annual-report-2019/ (Accessed: 20 March 2021)

Stone, M. (2020) 'While Microsoft Was Making Its Climate Pledge, It Was Sponsoring an Oil Conference,' *Vice*, 24 January [Online]. Available at: https://www.vice.com/en/article/xgqypn/while-microsoft-was-making-its-climate-pledge-it-was-sponsoring-an-oil-conference (Accessed: 20 March 2021)

Target (2012) *Community & Store Safety* [Online]. Target. Available at: https://corporate.target.com/corporate-responsibility/safety-preparedness/community-store-safety (Accessed: 20 March 2021)

Taylor, D.E. (2014) The State of Diversity in Environmental Organizations. Available at: https://orgs.law.harvard.edu/els/files/2014/02/FullReport_Green2.0_FINALReduced Size.pdf (Accessed: 20 March 2021)

Taylor, D.E. (2016) *Rise of the American Conservation Movement: Power, Privilege, and Environmental Protection*. Durham, NC: Duke University Press.

Taylor, D.E. (2018) *Diversity in Environmental Organizations Reporting and Transparency*. Available at: https://www.researchgate.net/publication/322698951_Diversity_in_Environmental_Organizations_Reporting_and_Transparency?channel=doi&linkId=5a69e2074585154d15450d78&showFulltext=true (Accessed: 16 June 2020)

The Washington Post (2019) Washington Post-Kaiser Family Foundation poll. Available at: https://games-cdn.washingtonpost.com/notes/prod/default/documents/fd042513-7ab8-40e0-9551-5f844587dbdd/note/4b15d814-7840-4c50-8e78-cf8ad6d5aeb0.pdf#page=1 (Accessed: 20 March 2021)

The World Bank (2017) Civil Society Policy Forum [Online]. The World Bank. Available at: https://www.worldbank.org/en/events/2017/04/21/civil-society-policy-forum (Accessed: 20 March 2021)

Uber (2020) *Let Me Start by Saying I Wish I Never Had to Send This Email* [Twitter]. 5 June. Available at: https://twitter.com/Uber/status/1268724613384515585 (Accessed: 20 March 2021)

UNFCC (2018) *Statistics on Participation and In-Session Engagement* [Online]. UNFCC. Available at: https://unfccc.int/process-and-meetings/parties-non-party-stakeholders/non-party-stakeholders/statistics-on-non-party-stakeholders/statistics-on-participation-and-in-session-engagement (Accessed: 20 March 2021)

UNICEF (2020) *More Than 1,100 Climate Actions Taken by Thousands of Adolescents from Latin America and the Caribbean* [Press release]. 4 June. Available at: https://www.unicef.org/lac/en/press-releases/more-1100-climate-actions-taken-thousands-adolescents-latin-america-and-caribbean (Accessed: 20 March 2021)

United Nations (2019) Civil Society [Online]. United Nations. Available at: https://metoomvmt.org/learn-more/statistics/ (Accessed: 20 March 2021)

US Senate, Committee on Energy and Natural Resources (1988) *Congressional Testimony of Dr. James Hansen*. Hearing. 23 June. Washington, DC: Government Printing Office. Available at: https://babel.hathitrust.org/cgi/pt?id=uc1.b5127807&view=1up&seq=45&skin=2021 (Accessed: 1 September 2021)

VanDyck, C.K. (2017) *Concept and Definition of Civil Society Sustainability*. Available at: https://csis-website-prod.s3.amazonaws.com/s3fs-public/publication/170630_VanDyck_CivilSocietySustainability_Web.pdf (Accessed: 20 March 2021)

Vaughan, A. (2014) 'Greenpeace Losses: Leaked Documents Reveal Extent of Financial Disarray,' *The Guardian*, 23 June [Online]. Available at: https://www.theguardian.com/environment/2014/jun/23/greenpeace-losses-financial-disarray (Accessed: 20 March 2021)

Wallman, D., Wells, E.C. and Rivera-Collazo, I.C. (2018) 'The Environmental Legacies of Colonialism in the Northern Neotropics: Introduction to the Special Issue', *Environmental Archeology*, 23(1) [online]. Available at: https://www.tandfonline.com/doi/full/10.1080/14614103.2017.1370857 (Accessed: 20 March 2021)

Walmart (2018) *Your Story Is Our Story. Culture Diversity & Inclusion*. Available at: https://corporate.walmart.com/media-library/document/2018-culture-diversity-inclusion-report/_proxyDocument?id=00000168-4df5-d71b-ad6b-4ffdbfa90001 (Accessed: 20 March 2021)

Walmart (2020) Making a Difference in Racial Equity: Walmart CEO Doug McMillon's Full Remarks [Online]. Walmart. Available at: https://corporate.walmart.com/equity (Accessed: 20 March 2021)

Weart, S. (2011) 'The Oxford Handbook of Climate Change and Society' in Dryzek, J.S., Norgaard, R.B. and Schlosberg, D. (eds.) *The Oxford Handbook of Climate Change and Society*, Oxford: Oxford University Press, pp.67–81.

Wheeler, S. (2020) 'UN Peacekeeping Has a Sexual Abuse Problem,' *Human Rights Watch*, 11 January [Online]. Available at: https://www.hrw.org/news/2020/01/11/un-peacekeeping-has-sexual-abuse-problem (Accessed: 20 March 2021)

World Economic Forum (2013) *The Future Role of Civil Society*. Available at: http://www3.weforum.org/docs/WEF_FutureRoleCivilSociety_Report_2013.pdf (Accessed: 20 March 2021)

World Economic Forum (2019) 'The Vital Role of Women in Creating a Lasting Afghan Peace,' *World Economic Forum*, 8 March [Online]. Available at: https://www.weforum.org/agenda/2019/03/why-women-need-to-be-included-in-peace-talks-and-what-happens-when-they-re-not/ (Accessed: 20 March 2021)

Xu, T. (2020) Standing Together for Justice [Blog] *DoorDash Blog*. Available at: https://blog.doordash.com/standing-together-for-justice-dc98cf164b7b (Accessed: 20 March 2021)

Younis, M. and Rzepa, A. (2019) One in Three Worldwide Lack Confidence in NGOs [Blog] *Gallup Blog*. Available at: https://news.gallup.com/opinion/gallup/258230/one-three-worldwide-lack-confidence-ngos.aspx (Accessed: 21 August 2020)

CONCLUSION

Vision 2030: building a low-carbon, climate-resilient, and inclusive world

This book made a case for a new corporate climate leadership. In concluding the book, we now make a case for a new corporate climate leader. Those reading this book will likely find themselves in a leadership position within a company or are training to become the leaders of tomorrow. What attributes does such a leader need? What will be the path for leading change within a company, across complex global supply chains, and within private sector collaborations?

Leading change in the climate-compatible company

Throughout their careers, the authors have been inspired and guided by theories of change developed by a number of academics specialized in policy and corporate transformation. The combined expertise of Elkington, Kotter, Kennedy, and Kingdon has been consolidated into four principles for guiding the new corporate climate leader and leading change in the climate-compatible company.

From the tactical to the transformational: a dynamic vision of leadership

John Elkington has written of his frustration working with leaders focused on making incremental changes to the status quo. Confronted with the nexus of pressing social and ecological problems at the dawn of the new decade, he has urged them to move from the tactical to the transformational and embracing the notion of "future quo."[1] The volume of companies currently committing to climate action is impressive, but the new entrants are now adopting a 2015 model of leadership rather than anticipating what leadership in 2030 will entail. The benchmark in 2015 was a science-based target consistent with holding temperatures below 2°C by the end of the century. Going forward, leadership will require *avoiding the unmanageable while managing the unavoidable*. In specific terms this means:

- A GHG reduction target consistent with holding global mean temperature rises to less than 1.5°C. This translates into a goal of net-zero emissions by 2050 and an interim target of 45% emissions reductions by 2030. It also requires implementing this absolute target across the full supply chain. Increasingly companies who adopt a conditional target (i.e. applying it to certain divisions or product lines) are being called out for inconsistency, and their reputations are taking a hit.
- A commitment to building resilience by enhancing capacity to anticipate, avoid, absorb, and recover from climate risk. This means investing in human, social, natural, physical, financial, and political capital inside the company, across the value chain, and within frontline communities.
- Building an inclusive economy. Being "less intrusive on communities" will no longer be sufficient. It will be necessary to explore how the company can work to create a shared prosperity and a "just transition," not just for the communities who are currently dependent on the high-carbon economy, but for those who will build the low-carbon and resilient economy of the 21st century.

Companies typically need a prompt to push them to think beyond the horizon. This could be pressure from an NGO, the result of shareholder activism, the consequence of a shifting regulatory environment, competitive pressures from a rival, understanding of climate risk, consumer demand, or sensing an opening. For example, many companies embark on a climate leadership journey following pressure to disclose from a peer company. Others, rebound from reputational damage following scrutiny from an advocacy group. Still others begin their climate journey when they suffer supply chain disruption or assess that the regulatory environment is creating heightened transition risk. An enlightened group of companies begin to move when they see opportunities emerge – perhaps the shifting investment patterns of millennials, perhaps changing consumer preferences, maybe the onset of significant amounts of public money, or occasionally the emergence of a game-changing innovation. John Kingdom describes these pressures that elevate climate change on the decision agenda and offer alternatives to existing policies and business models as "windows of opportunity." This is when separate streams of problems, solutions, and politics converge to move an issue.[2] The new corporate climate leader needs to recognize these separate streams and elevate them to the C-suite. This book offers the balance between risk, reward and response precisely because these are the necessary elements to ensure that corporate leadership is aware of the threat to the status quo, sees value in the future quo, and is confident that the transition from here to there can be achieved. One way of enhancing that confidence is to generate short-term wins. Short-term operational and reputational rewards embed the strategy and encourage leadership to press ahead. This could be as simple as receiving positive press, or even offsetting criticism of other parts of the company. Over the years, Walmart has received a lot of positive coverage for its approach to climate change, and this has often helped offset its more questionable record on the social side of sustainability. The surest way to sustain investment in transformation is to engineer stretch targets and sequencing when making climate commitments. This book has argued that long-term goals – a net-zero target by 2050 – without an interim target undermine integrity and credibility. A more positive twist is that the presence of a short-term or intermediate goal, and the achievement of that goal, enable companies to grow in confidence, learn, and then feel that stretch goals are reachable.

Imagination and not just ambition

Those of us who have spent years working on climate change are accustomed to hearing one word again and again – "ambition." Governments are constantly being pressed to raise the ambition of their economy-wide GHG reductions targets. They are asked to increase the ambition within their energy policy, or their overseas development assistance. Companies are pressed to raise the ambition of their corporate commitments. Even philanthropies and civil society organizations are pushed to raise the ambition of their spending or their advocacy. Increased ambition is certainly needed, particularly as we embark on this decisive decade, but what if an unrelenting and narrow focus on ambition is preventing the climate community from embracing the one essential attribute vital to solving the climate crisis? This attribute is imagination. Consider these examples. For years, climate advocacy groups in the United States invested all their time, money, energy, and political capital into a push for rejoining the Paris Agreement while ignoring the ways in which President Trump was fundamentally changing the composition of the Federal judiciary, a move that has more damaging and long-lasting consequences than withdrawal from the Paris Agreement. When the COVID pandemic struck, these same groups spent all their time preparing for the next round of climate negotiations under the United Nations Framework Convention on Climate Change at COP26 in Glasgow and largely neglected the opportunity to shape more than $15 trillion dollars in economic recovery and stimulus spending across the G20. In 2021, these same organizations are focused on updating the US national climate action plan while ignoring the voting rights legislation stalled in the US Congress – a measure that might do more to lay the foundations of a low-carbon and resilient transition in the United States than any energy policy. The list of examples is unfortunately a long one. For years the climate community has ignored the transport and land-use sectors by focusing too rigidly on energy. Equally, we have failed to move beyond emissions reductions to give the same weight and attention to resilience and finance. Most concerning of all, the community has tended to think of climate change as an environmental problem in search of an environmental policy solution, and therefore has ignored the full suite of economic, cultural, and social transformation necessary to really address this crisis. Similar patterns exist in the world of corporate climate leadership, as revealed in this book. Food companies, which may have women constituting up to 80% of their supply chain workforce, think of climate risk as salt-water intrusion to be tackled by flood defenses and ignore the role gender discrimination and broader inequality plays in amplifying risk. Banks make commitments to reduce their energy use and shift their energy mix while pouring billions into fossil fuels. These are all failures of imagination – they reflect an inability to look beyond our silos and to see the bigger web of how climate risk connects to other global challenges; and how the solution to the climate crisis may involve innovations beyond climate policy. Therefore, as we move through this decade with a renewed commitment to greater ambition, we must also inject a far greater level of imagination into our thinking.

Cathedral thinking – the vision to launch multigenerational and multistakeholder initiatives

The benchmark of climate leadership is a net-zero goal achieved by 2050. Given that most CEOs remain in post for an average of six years this means instituting change that will need to be sustained over a long period of time.

The first step is to create a credible process that can ensure broad-based support across the company. This typically begins by creating a baseline – a credible assessment of the company's emissions and climate risk. The second step usually involves benchmarking – understanding what peer companies are doing, assessing where the laggards sit, and identifying where those in the vanguard are doing and how feasible it will be to join them as pioneers of climate leadership. Sustainability departments often kick-start the process, but they are often marginalized within companies and therefore a more powerful leader or series of leaders needs to be found to push the strategy through. The most successful companies achieve a distributed form of leadership, whereby every member of the C-Suite including the Chief Financial Officer, Chief Technology Officer, Chief Procurement Officer, and so on, have climate-related performance targets of their own. Companies often suffer setbacks if the passion and drive for climate leadership come exclusively from one individual. To avoid this "key person risk" it is important to enlist a volunteer army inside the company and within strategic partners (i.e. across the supply chain, within business federations, and within government) to aid the achievement of the long-term goal.[3]

Changing the terms of the debate can be vital for building a coalition. One of the great failures of the environmental community has been a failure to translate risk and rewards into language that resonates with corporate leaders and speaks to their incentives and motivations. This has begun to change in recent years, but the moral case for climate leadership still crowds out a tailored business case – made specific for different sectors, different companies, and even for unique individuals in specific departments. Creating what Kennedy calls a vocabulary of arguments[4] – a combination of the right words, data points, and incentives – to appeal to different audiences inside the company is vital. This in turn helps with "foregrounding," the need to ensure that climate change remains high on the corporate agenda, even when confronted with other material business risks or priorities.[5] Too often, companies focus on climate change when there is an external forcing mechanism – a major political summit, the release of the latest IPCC report, the threat of shareholder activism at an AGM – but lose momentum during quieter periods. Addressing climate change requires both urgency and persistence, and so "foregrounding" is essential.

Courage and patience in a time of urgency

Years ago, the authors were influenced by a book called *The Impossible Will Take a Little While: A Citizen's Guide to Hope in a Time of Fear*. A collection of vignettes from a variety of change agents ranging from a woman servicing a suicide hotline to Nelson Mandela, this book speaks to the need to preserve a sense of patience and personal resilience even when confronted with urgency.[6] The new corporate leader will need to combine patience with urgency – no small task. They will need to understand that building coalitions takes time; persuading colleagues and superiors to lead requires significant investments of labor and considerable persuasive ability; and that it is only ultimately worth it if a sequence is set that leads to transformation.

In addition, courage will be the key attribute for the new corporate climate leader – the courage to speak hard truths to colleagues and also the courage to get out of professional silos and comfort zones. There is an old adage that few conflicts are solved without engaging the combatants. Similarly, it is impossible to properly manage and ultimately solve the climate crisis without engaging and working with those countries and companies who are driving the crisis. Both authors advocate for and encourage engagement. On the other hand, there is a fine line between engaging countries or companies and being coopted by them. In recent years, there has been a worrying tendency to cross that line. Today, companies who fail to pay their taxes, and consequently starve the public purse of the vital financial resources necessary for decarbonization and resilience building, are called climate leaders and provided with privileged positions to fund and lead climate advocacy. Companies with the most appalling labor relations and workers' rights records are lauded because they have ambitious emissions reductions plans. Other companies are given a free pass when announcing climate commitments that are clearly public relations exercises, in large part because they are channeling large amounts of money into climate think tanks and nonprofit organizations. And the problem extends to countries too. There has been an unfortunate tendency to turn a blind eye to dreadful human rights abuses within countries so long as their climate goals seem appropriately ambitious. Those of us working on climate change need to find the wisdom to engage with one another constructively but retain the courage to speak and hear the truth about our efforts.

This book has argued that the new corporate climate leadership needs to be *dynamic and robust* because the current climate leadership is not suited to address the pace, scale, and breadth of issues related to the climate crisis. What we present is also a case for leadership, because it is clearly goal-oriented, it will leverage the efforts of all toward this goal, and it will take courage. And it is, finally, a case for climate action understood much more broadly than it is today. From the time in which it has entered public discourse in the 1980s to the time it became a mainstream topic thanks to Al Gore's "An Inconvenient Truth," our common understanding of climate action has been limited to the sorts of actions that science dictates, mostly, that is, reducing GHG emissions.

There is no doubt that this is a necessary – and perhaps the first – step in addressing the root physical cause of climate impacts. We have argued for this time and time again in this volume. But it is not a sufficient one. It is not sufficient because, as we have also argued, the causes of climate impacts are of many different kinds, and not only physical. Hence addressing the climate crisis will draw on actions that touch upon those causes too: social, economic, political, financial, and cultural.

To rethink climate leadership, then, is to rethink climate. This is what we hope to have achieved here. Climate, it is often said, lies at the nexus of issues, and only an approach that tackles this can meet the challenge at the right pace and at the right scale. This explains the complexity of our climate crisis and the somewhat common impression that it is a problem too large for a single actor to solve. It also explains why, in spite of what we have argued are for the most part genuine – and impressive – attempts at contributing to this solution, the private sector as so far fallen short.

To rethink climate then is to rethink leadership. It means expanding the common understanding of the purview of business action to include social, economic, political, financial, and cultural arenas too. This reshapes business goals, partnerships, the time horizon companies view themselves as operating in, and the role of business in the world beyond profit-maximizing. It takes new forms of courage for individual leaders who, backed by science and facts of the kinds we have presented here, can blaze trails into these new territories. Some companies have, it is true, already started exploring in this direction, but we have yet to see sufficient numbers of companies do so with sufficient integrity, ambition, and breadth to collectively accelerate the transition to a low-carbon, resilient and inclusive future.

Beyond addressing climate change as a threat, we also highlighted that a more integrated approach to it opens up opportunities far richer than carbon emissions reductions. By addressing these multiple facets, companies can proactively shape the future we are headed toward instead of reactively averting – or hoping we will avert – the worst impacts of climate change. We can collectively reshape how we think of labor, gender relations, our relationship to the so-called natural world, how and what we consume, our health, the cities and environments we dwell in, and our relationship with frontline and marginalized communities.

The new climate leadership, then, is about building the world we want by redefining values and, ultimately, redefining, value. What are the attributes that make a leader suited to face the climate crisis? What values must a company have in order to be orientated toward the complex goal of addressing climate change systemically? Moreover, what value can a company be understood as creating once it is understood as creating more than economic value? Indeed, making a case for the new climate leadership amounts to making a case for a new definition of value creation altogether.

We hope to have achieved two things with this book: to have charted the path for this work of redefining the role of the private sector and the kind of value it can see itself as creating, and to have taken a first step on that path. We recognize that the path will not be linear; it may be arduous, it may even seem impossible at times. But we believe, based on our research and experience, that it is possible; it is also urgent, rewarding, and the opportunity of a generation if not of a century.

TS Eliot wrote: "for last year's words belong to last year's language. And next year's words await another voice. And to make an end is to make a beginning." We have made an end to this book. We hope we have made a beginning in clearing a path for the next generation of corporate climate leaders. We look forward to hearing their voices as they champion the cause of a just and sustainable world.

References

1. Elkington, J. (2020) *Green Swans: The Coming Boom in Regenerative Capitalism*. New York: Fast Company Press.
2. Kingdon, J.W. (1984) *Agendas, Alternatives, and Public Policies*. Boston: Little, Brown and Company.
3. Kotter, J. (2012) *Leading Change*. Cambridge, MA: Harvard Business Review Press.
4. Kennedy, D. (2005) "Challenging Expert Rule: The Politics of Global Governance". *Sydney Law Review 27*: 5–15.
5. Giddens, A. (2009) *The Politics of Climate Change*. Cambridge: Polity Press.
6. Loeb, P.L. (2014) *The Impossible Will Take a Little While: A Citizen's Guide to Hope in a Time of Fear*. London: Basic Books; 2nd edition.

INDEX

Note: Bold page numbers refer to tables; Page numbers followed by "n" refer to notes.

AB InBev 70
Abrams, F. 53–54
adaptation finance and investing, opportunities for leadership through 161–162
Adaptation Fund 161
Adidas 117
AECOM 145
Africa 197
African Development Bank 186
African Youth Initiative on Climate Change 37
agency theory 173
agriculture 76–81, 90, 114–117; decarbonization of financial flows 116–117; energy and transport commitments 115; opportunity related to economic goals 78–79; opportunity related to environmental goals 79–80; opportunity related to social goals 80–81; opportunity related to strategic goals 79; resilience commitments 117; value chains 90–92
agroecology 80
agroforestry 80
Alphabet 63
Alpha Natural 31
Amazon 39, 63, 115, 200–201; Climate Pledge 119; effective tax rate 145–146; stranded assets 157

American Petroleum Institute 190
AMOC *see* Atlantic Meridional Overturning Circulation (AMOC)
Apollo Program 43
Apple 43; inconsistency commitment 119; stranded assets 157; Supplier Clean Energy Program 69
Arab Spring 188
Arch Coal 31
Asia 5, 23
assets 152–163; stranded 156–157
assets under management (AUM) 67, 111, 116, 160, 191
AstraZeneca 76
Atlantic Meridional Overturning Circulation (AMOC) 5–6
AUM *see* assets under management (AUM)
Australia: bush fires in 158; climate-related disasters 24; electricity generation 70; GDP 46; GHG emissions 46

Bahrain 188
Bakan, J. 146
Baker, E. 203n1
Baker, J. 120
Banerjee, A. 45
Bank of America 55; contradictions with business goals 122

218 Index

Bank of England 162; Prudential Regulation Authority 154
Barclays 116; contradictions with business goals 122
BCI 117
Benefit–Cost ratios 100, 111, 141
Benioff, M. 55
Ben & Jerry 55
Better Cotton Principles and Criteria 117
Bezos, J. 39, 200
Biden Administration 110–111
Black Lives Matter (BLM) 199, 200
BlackRock Inc. 2, 24, 54, 66, 116, 153, 201
BLM *see* Black Lives Matter (BLM)
Bloomberg 66
Bloomberg, M. 195
BMW Group 75
BNP Paribas 116, 161
BNY Mellon 140
BP 145, 201; inconsistency commitment 119; lobbying 180; stranded assets 157
Brazil 134
British Academy: *Principles for Purposeful Business* 54
BSR *see* Business for Social Responsibility (BSR)
Burke, M. 152
Burke, T. 187
business as usual 4, 24, 63, 102, 119, 131, 153, 159, 201, 202
Business for Social Responsibility (BSR) 133, 139
business goals undermining climate leadership, contradictions with 120–122
business leadership, civil society and 198–202
Business Roundtable 145, 168; *Statement on the Purpose of a Corporation* 54

California Public Employees' Retirement System (CalPERS) 29
California State Teachers' Retirement System (CalSTRS) 139
CalPERS *see* California Public Employees' Retirement System (CalPERS)
CalSTRS *see* California State Teachers' Retirement System (CalSTRS)
Cambodia 134
CAN *see* Climate Action Network (CAN)
Canada 139; transition in energy systems 26

capital: assets 13, 95; financial 99–100; human 96; natural 97–98; physical 98–99; political 100–101; social 96–97
Carbon Dioxide Removal (CDR) 91, 114
Carbon Disclosure Project (CDP) 133, 191
carbon pricing 163
Carbon Pricing Leadership Coalition 144
carbon removal 7
carbon sink 114
Cargill 114
cathedral thinking 213–214
CDC *see* Centers for Disease Control (CDC)
CDP *see* Carbon Disclosure Project (CDP)
CDR *see* Carbon Dioxide Removal (CDR)
CEEF *see* Commercializing Energy Efficiency Finance (CEEF)
Centers for Disease Control (CDC) 5
CEO activism and leadership 55–56
Chanel 39
Chevron 153; lobbying 180
China 23, 139, 144, 194; China Utility-Based Energy Efficiency Finance Program 161; climate-related disasters 24; contradictions with business goals 120, 121; electricity generation 68; environmental education 38; GHG emissions 46; transition in energy systems 26
Citi: contradictions with business goals 122
Civil Rights movement 187, 203n1
civil society: and business leadership 198–202; and climate action 188–202; in context 186–188; definition of 186; inauthenticity 194–197; inconsistency 194–197; leadership structures 192–194; reciprocally co-opting 201–202
civil society organizations (CSOs) 186, 188, 190, 191, 195, 203
Clark, S.P. 203n1
Client Earth 119–120
climate action 1, 2, 4, 10–13, 24, 25, 37–39, 52, 56, 63–67, 73, 81, 110, 111, 131, 139, 144, 152, 153–156, 159, 175–181, 185–188, 193, 195, 196, 210, 212, 213, 215; beyond climate 162–163; civil society and 188–192, 198–202; standard theories of barriers to 40–47
Climate Action 100+, 66, 153
Climate Action Network (CAN) 190

Index **219**

Climate Ambition Alliance 118
climate change 21–32, 189; corporate climate risk 26–31; physical risks from 21–24, 155; politics 38; transition risks from 24–26, 155–156, 158
climate-compatible company, leading change in: dynamic vision of leadership 210–212; imagination and ambition 212–213
climate-compatible economy, building 89–105; decarbonization 89–95; just transition, securing 101–105; socio-ecological resilience, enhancing 95–101
Climate Group 76
climate injustice 193
climate leadership, corporate governance and 175–176
climate litigation 37–38
climate opportunities, corporate governance and 176–178
climate-related disasters 1–2, 22, 23
climate-related opportunities for finance 159–162; opportunities for leadership through adaptation finance and investing 161–162; opportunities for leadership through changing lending policies 160–161; opportunities for leadership through "sustainable finance" 159–160
climate-related risks to finance 155–159, **158**
climate risk, diagnosing 133–135
Coal India 103
Coalition of Jakarta Residents Opposing Water Privatization 186
Coates, T.-N. 3
Coca Cola Company, The 55
Cohen, B. 55
Colombia 134
Commercializing Energy Efficiency Finance (CEEF) 161
compliance and legal risk 28
confusing power 199–201
Conway, E.: *Merchants of Doubt* 190
COP21, 53
Copenhagen Accord (2009) 4, 131
corporate climate commitments: agriculture and land use commitments 114–117; inconsistencies and contradictions 118–122; net-zero goal 112–114; science-based targets 112–114
corporate climate leadership 13, 111, 112, 117, 118, 120, 122; future 132; new 11, 12, 14, 89, 123, 147, 172, 210, 215

corporate climate risk 26–31; compliance and legal risk 28; financial risk 28–29; human resources risk 27–28; operational and supply chain risk 27; profiles in 30–31
corporate commitment to resilience 136–139
corporate communications about climate 39
corporate governance: and climate leadership 175–176; and climate opportunities 176–178; purpose and 173–175, **175**; and risk management 176–178; and transformation of the economy toward low emissions or net zero 178–181
corporate social responsibility (CSR) 10, 12, 47–49, 64, **65**
Cote d'Ivoire 134
CO_2 emissions 6, 39–40, 45–47, 67, 71, 90, 91, 93, 94, 118, 188, 196
courage 214–216
Crédit Agricole 116
critical theory framework 3
CSOs *see* civil society organizations (CSOs)
CSR *see* corporate social responsibility (CSR)
C-suite 48, 211, 213
Cullors, P. 187

DACS *see* direct air capture and storage (DACS)
Daimler 75, 177–178
Dauvergne 120
Davos Manifesto 2020: the Universal Purpose of a Company in the Fourth Industrial Revolution, The (WEF) 54
de astries, H. 29
decarbonization 89–95, 144, 155, 158; enabling in others 139–144; of financial flows 116–118; food 90–92; footnotes 94–95; fuel 92–94; inside the new corporate climate leader 133–139
Dell, M. 152
Denmark: environmental education 38
deprivation: conflicts 1; economic 1
Diaz, D.B. 152
direct air capture and storage (DACS) 7, 94
documentary analysis 3
DoorDash 200
Dow Chemicals 55
due diligence: on the company 198–199; on organizations 198

Duflo, E. 45
Duke's Fuqua School of Business 38
dynamic vision of leadership 210–212

EACs *see* energy attribute certificates (EACs)
East Asia 23
EAT–Lancet report (2019) 81
ecological marginalization 1
economic deprivation 1
economic goals **65**; agriculture 78–79; electricity generation 68–69; energy efficiency 72; heat generation 71; land use 78–79; transport 74
EDF *see* Environmental Defense Fund (EDF)
Egypt 188
EIB *see* European Investment Bank (EIB)
electricity generation: context 67–68; opportunity related to economic goals 68–69; opportunity related to environmental goals 69; opportunity related to social goals; opportunity related to strategic goals 69
Eliot, T.S. 216
Elkington, J. 47–51, 145
Emissions Gap Report of 2011, 4, 131
employee activism 38–39
energy and transport commitments 115
energy attribute certificates (EACs) 69
energy demand 45–47
energy efficiency: opportunity related to economic goals 72; opportunity related to environmental goals 72; opportunity related to social goals 72–73; opportunity related to strategic goals 72
energy systems: electricity generation 67–70; energy efficiency 72–73; heat generation 70–71
entrepreneurship 42
Environmental Defense Fund (EDF) 189, 194, 195
environmental education 38
environmental goals **65**; agriculture 79–80; electricity generation 69; energy efficiency 72; land use 79–80; transport 76
environmental injustice 193
Environmental Profit & Loss (EP&L) 39
EP&L *see* Environmental Profit & Loss (EP&L)
ESG market 159–160
ETF *see* exchange-traded fund (ETF)

ethical behavior 171
Ethiopia 134
Europe 4, 139, 190; transition in energy systems 26
European Commission 21
European Investment Bank (EIB) 139
European Union (EU) 24, 91; Green Deal 25
exchange-traded fund (ETF) 66
Extinction Rebellion (XR) 37
extractivism 40–42
extreme events 5
Exxon Mobil 116, 201; lobbying 180; stranded assets 157

Facebook 115, 188
Fair Tax Mark 145
FAO *see* Food and Agriculture Organization (FAO)
Federal Reserve 21
Federal Reserve Bank of Richmond 152–153
Ferrero 134
Fiat Chrysler 75, 76, 115
financial capital 99–100
financial flows: decarbonization of 116–117; toward decarbonization and resilience, shifting 139–141
financial rewards 152–163; assets and liabilities 152–163; climate action beyond climate 162–163; climate-related risks to finance 155–159
financial risk 28–29
Financial Stability Board (FSB) 1, 11, 21; Task Force on Climate-related Financial Disclosures (TCFD) 66
Fink, L. 54, 153
fiscal accountability 179
Floyd, G. 188, 200
FOLU *see* Food and Land Use Coalition (FOLU)
food, reducing emissions across 90–92
Food and Agriculture Organization (FAO) 4, 92, 194
Food and Land Use Coalition (FOLU) 78
food insecurity 4
food security 4
footnotes: decarbonization 94–95
Ford 75
Ford Chrysler 190
foregrounding 214
Forest Positive Coalition of Action 114
fossil fuel industries 25–26
France 147

Francis, C. 53
Frazier, K. 55
Friedman, M. 54
Friends of the Earth 190
FSB *see* Financial Stability Board (FSB)
FSG Social Impact Advisors 51
fuel: decarbonization 92–94
Fujitsu 70

Gallup 188
Garza, A. 187
GCC *see* Global Climate Coalition (GCC)
GDP *see* gross domestic product (GDP)
gender diversity 194
gender equality 7–9
Genentech 76
General Foods 53
General Mills 114
General Motors 75, 115
Generation Investment Management 195
generation Z 38
Ghana 134
GHG *see* global greenhouse gas (GHG) emissions
Global Climate Coalition (GCC) 180, 190
Global Commission on Adaptation 100, 141
global greenhouse gas (GHG) emissions 3–4, 6, 7, 25, 39, 41, 49, 76, 77, 89–91, 93, 94, 113, 114, 116, 131, 136, 191, 215
Global Investor Coalition on Climate Change 29
Global Investor Statement to Governments on Climate Change 146
Global Organic Textile Standard (GOTS) 117
global warming 189
Global Warming in an Unequal World (Centre for Science and Environment) 190
Global Warming: The Greenpeace Report (Greenpeace) 190
Goldman Sachs 55, 116, 195
Google 115; algorithmic search engine 42
Gore, A. 195, 215
GOTS *see* Global Organic Textile Standard (GOTS)
green bond issuance 66
Greenfield, J. 55
greenhouse effect 189
Greenhouse Gas Protocol 118
green industrial revolution 25
Green New Deal 38, 197

Greenpeace 195–196; *Global Warming: The Greenpeace Report* 190
Greenpeace International 189
green swans 49–50
greenwashing 199; accusations, inconsistent commitments and 118–120
gross domestic product (GDP) 9, 26, 29, 30, 41, 50, 54, 67, 140, 143, 144, 146, 154, 179; fetishism 43–47, 50
G20, 1, 21, 212

Hamer, F.L. 203n1
Hansen, J. 9
health stressors 6
heat generation: context 70–71; opportunity related to economic goals 71; opportunity related to social goals 71; opportunity related to strategic goals 71; Hewlett-Packard 2; climate risk 30
higher-order business goals **65**
H&M (formerly Hennes & Mauritz) 117
Honda 75
HSBC 116, 161; Climate Vulnerability Assessment 162
Hsiang, S.M. 152
human capital 96
Human Development Report 2019, 101
human resources risk 27–28
human rights, strengthening 141–144
hunger 6
Hurricanes: Florence 2, 24; Harvey 2; Irma 2, 24; Maria 30–31; Michael 2; Sandy 30

IEA *see* International Energy Agency (IEA)
IFRC *see* International Federation of Red Cross and Red Crescent Societies (IFRC)
Ikea 117
ILO *see* International Labour Organization (ILO)
IMF 179
Impossible Will Take a Little While: A Citizen's Guide to Hope in a Time of Fear, The 214
inconsistent commitments, and greenwashing accusations 118–120
India 46; Centre for Science and Environment 190
Indonesia 134
infrastructure risk, climate change and 27

ING Group 161
innovation agenda 131–147
Intergovernmental Panel on Climate Change (IPCC) 3, 6, 7, 10, 22, 76, 95, 136, 180, 189, 214; agriculture and land use commitments 114; AR5 report 73; climate risk, definition of 137, 138; global food system 90; resilience, definition of 95, 138
International Council for Clean Transportation 74
International Covenant on Economic, Social and Cultural Rights: Article 12, 9
International Energy Agency (IEA) 45–46, 71, 121
International Federation of Red Cross and Red Crescent Societies (IFRC) 5
International Labour Office 143
International Labour Organization (ILO) 8, 101–102, 142
International Organization of Motor Vehicle Manufacturers (OICA) 120
International Renewable Energy Agency (IRENA) 68, 70, 140
International Resource Panel 41
IPCC *see* Intergovernmental Panel on Climate Change (IPCC)
IRENA *see* International Renewable Energy Agency (IRENA)

Japan 139; transition in energy systems 26
Johnson, L.B. 2
Johnson, M.P. 187
Jones, V.M. 203n1
JPMorgan Chase 31, 55; contradictions with business goals 121
"just-in-time" supply chains 31
just transition, securing 101–105

Kaysen, C. 53
Keeling, C.D. 2
KPMG 48
Kramer, M. 51
Kyoto Protocol 190

labor relations 179–180
Lancet Commission 1, 5
land use 76–81, 90, 114–117; decarbonization of financial flows 116–117; energy and transport commitments 115; opportunity related to economic goals 78–79; opportunity related to environmental goals 79–80; opportunity related to social goals 80–81; opportunity related to strategic goals 79; resilience commitments 117
Law Enforcement Grant Program 200
Lawrence Livermore Laboratory 93
leadership: CEO 55–56; corporate climate 11; current 9–11; structures 192–194
Learning Groups 117
LeasePlan 74
Levi Strauss 145
liabilities 152–163
Libya 188
literature analysis 3
lobbying 180–181
location risk, climate change and 27
logistics risk, climate change and 27
Lyft 200

Mandela, N. 214
market improvement, accountability for climate-related opportunities for finance 159–162; mandatory climate disclosure to increase transparency 146–147; policy enabling environment, shaping 144–146
Mars 70, 134
Mawdsleys 76
Mazzucato, M. 42, 43
McDonald's 115
Mendocino Complex Fire of 2018, 24
Mercedes Benz 177–178
Merchants of Doubt (Oreskes and Conway) 190
MeToo movement 187
Microsoft: stranded assets 157
Miguel, E. 152
MIT's Sloan School of Management 38
Mondelez International 134
Montgomery Bus Boycott 187
Moore, F.C. 152
morality 171
Morgan, J. 122
Morgan Stanley 29, 116, 153
Morocco 188
Moynihan, B. 55–56
MSCI ACWI ESG Leaders Index 160
MSCI All-Country World Index 160
Munich Re 158
Musk, E. 42

NAFTA 180
National Institute of Building Sciences 99–100, 111, 141

Nationally Determined Contributions (NDCs) 39
National Oceanic and Atmospheric Administration 153
natural capital 97–98
natural disasters 1, 22
NDCs *see* Nationally Determined Contributions (NDCs)
Nestlé 134
net-zero goal 112–114
net-zero greenhouse gas emissions, by 2050, 6–7, 11
new climate economy, building 63–81; energy systems 67–73; land use and agriculture 76–81; transport 73–76
NewClimate Institute 113, 118, 136
New Economics Foundation 23
New York Times 190
New Zealand 147
Nike 117
Nipponkoa, S.J. 99
Nissan 75
Nixon, R. 189
Nordhaus, W. 152
North America 92
Northern America 4, 7
Novo 76
NRDC 194–195

Ocasio-Cortez, A. 197
OCS *see* Organic Content Standard (OCS)
OECD *see* Organisation for Economic Co-operation and Development (OECD)
Office for the United Nations Commissioner for Human Rights (OHCHR) 9
OHCHR *see* Office for the United Nations Commissioner for Human Rights (OHCHR)
OICA *see* International Organization of Motor Vehicle Manufacturers (OICA)
operational and supply chain risk 27
Oreskes, N.: *Merchants of Doubt* 190
Organic Content Standard (OCS) 117
Organisation for Economic Co-operation and Development (OECD) 142
Oxfam 186

Paris Agreement (2015) 2, 4, 10, 24, 53, 55, 118, 121, 122, 131, 132, 153, 156, 191; Article 4, 137
Parks, R. 187

participant observation 3
participatory decision-making 179–180
Patagonia 117
patience 214–216
Peabody 31
PG&E 2; utility bankruptcy 31
physical capital 98–99
physical risks from climate change 21–24, 155
policy enabling environment, shaping 144–146
policy engagement 52–53
political capital 100–101
Polman, P. 55
Porter, M.: Five Forces 51
poverty 6; multidimensional 101
powerful gestures 199–201
PPA 68
practitioner interviewing 3
Principles for Purposeful Business (British Academy) 54
private sector, myth of 42–43
procedural rights 7
Proctor & Gamble (P&G): contradictions with business goals 121
production and procurement shortfall risk, climate change and 27
progressive lobbying 52–53
Project Drawdown 6
Public Safety Focused Grant 200
Puerto Rico: Hurricane Maria impact in 30–31
purpose 169–170, 172–173; and corporate governance 173–175, **175**

Race-to-Zero campaign 113
Raworth, K. 122
RBS *see* Royal Bank of Scotland (RBS)
RE100 69, 70, 115
REBA *see* Renewable Energy Buyers Alliance (REBA)
reciprocally co-opting civil society 201–202
referral sampling 3
regenerative capitalism 49–50
Remote Sensing-based Information and Insurance for Crops in Emerging Economies (RIICE) 92
Renewable Energy Buyers Alliance (REBA) 115
resilience: definition of 95, 139; enabling in others 139–144; inside the new corporate climate leader 133–139; socio-ecological 95–101

224 Index

resilience commitments 117
resilience dividend 111, 141
resilient economy 9
Responsible Blockchain Sourcing Network 76
Return on Investment 67
Revelle, R. 2
RIICE *see* Remote Sensing-based Information and Insurance for Crops in Emerging Economies (RIICE)
Rio Earth Summit (1992) 189, 190
risk management, corporate governance and 176–178
robust commitment to net-zero emissions 135–136
Rolling Stone 131
Royal Bank of Scotland (RBS) 116, 139
Royal Dutch Shell: lobbying 180

S&P 500, 43
Santos 201
SBTs *see* science-based targets (SBTs)
science-based targets (SBTs) 112–114
self-interest 64
Sen, A. 44–45, 122
SFC *see* Smart Freight Centre (SFC)
shared value 51–52; definition of 51; pillars of 51
Shell 201; inconsistency commitment 119; stranded assets 157
SIGI *see* Social Institutions and Gender Index (SIGI)
SLM *see* sustainable land management (SLM)
slow violence of unsustainability 4–6
Smart Freight Centre (SFC): GLEC Framework 94
snowball sampling 3
social capital 96–97
social finance 140
social goals **65**; agriculture 80–81; electricity generation 69–70; energy efficiency 72–73; heat generation 71; land use 80–81; transport 76
Social Institutions and Gender Index (SIGI) 142
socio-ecological resilience, enhancing 95–101; financial capital 99–100; human capital 96; natural capital 97–98; physical capital 98–99; political capital 100–101; social capital 96–97
SolarCity 42
Solheim, E. 196
South Asia 23, 197

SpaceX 42
stakeholders 168, 172
stakeholder theory 173
Starbucks 145
Statement on the Purpose of a Corporation (Business Roundtable) 54
stewardship theory 173
Steyer, T. 195
Stiglitz, J. 44, 45
stranded assets 156–157
strategic goals **65**; agriculture 79; electricity generation 69; energy efficiency 72; heat generation 71; land use 79; transport 75
Sub-Saharan Africa 197
Sustainable Development Goals 122
sustainable finance, opportunities for leadership through 159–160
sustainable land management (SLM) 91
Swiss Re 154
Syria 188
systemic barriers, overcoming 37–56; CEO activism and leadership 55–56; corporate social responsibility 47–49; extractivism 40–42; GDP fetishism 43–47; policy engagement 52–53; private sector, myth of 42–43; progressive lobbying 52–53; purpose 53–55; regenerative capitalism 49–50; shared value 51–52

talent management 27, 28
Task Force on Climate-related Financial Disclosures (TCFD) 66, 117, 146–147, 163
TB Bank: contradictions with business goals 122
TCFD *see* Task Force on Climate-related Financial Disclosures (TCFD)
TCO *see* total cost of ownership (TCO)
Tesla 42
Textile Exchange: Organic Content Standard (OCS) 117
Thailand: corporate climate risk 30
Thatcher, M. 189
third sector 186
Thunberg, G. 37
Tometi, O. 187
Total: lobbying 180
total cost of ownership (TCO) 74
Toyota 75, 115
transition risks from climate change 24–26, 155–156, 158
transparency 179

transport 73–76; opportunity related to economic goals 74; opportunity related to environmental goals 76; opportunity related to social goals 76; opportunity related to strategic goals 75
Triangular Slave Trade 40–41
triple bottom line 49
Trump, D. 55, 145, 212
Tunisia 188
21st Conference of the Parties 191
Twitter 188

Uber 200
UBS Investor Watch Global Survey 160
UK FTSE 10
UK *see* United Kingdom (UK)
UN *see* United Nations (UN)
UNDP *see* United Nations Development Programme (UNDP)
UN Environment Program: Global Resources Outlook 41
UNEP *see* United Nations Environment Programme (UNEP)
UNFCCC *see* United Nations Framework Convention on Climate Change (UNFCCC)
UN Global Compact 160
UNICEF: Development Initiatives 4–5
Unilever 70; contradictions with business goals 121
United Kingdom (UK) 54, 147, 196–197; environmental education 38; Met Office 162
United Nations (UN) 6, 21, 190, 194, 196; Climate Change 144; COP 196; Food and Agriculture Organization 194; Principles of Responsible Investing 160; Sustainable Development Goals 122
United Nations Development Programme (UNDP) 101, 102; Human Development Report 144
United Nations Environment Programme (UNEP) 3, 6, 131, 141; Emissions Gap report for 2019, 39
United Nations Framework Convention on Climate Change (UNFCCC) 161, 189, 212
United States (US) 54, 139, 143, 145; Apollo Program 43; assets under management 111; carbon removal 7; CEO activism and leadership 55; climate and infrastructure spending 110–111; climate-related disasters 1–2, 23–24; Commodities Futures Trading Commission 147; Commodity Futures Trading Commission 24, 139; compliance and legal risk 28; Congress 153, 189, 212; conversation movement, emergence of 193; corporate climate risk 31; Department of Energy 42; electricity generation 68, 70, 93; Energy Information Agency 73; ETF 66; extreme events 5; footnotes 95; Green New Deal 38, 197; health implications 5; Justice Department 203n1; National Science Foundation 42; transition in energy systems 26; transition risks from climate change 25; transport 74; Union for Concerned Scientists 155
Universal Declaration of Human Rights (1948) 7; Article 12, 9; Article 25, 9
University of Vermont: MBA in Sustainable Innovation 38
US *see* United States (US)

Vale 98
value-chain management 91
Verizon 2, 115, 133–134; climate risk 30
Vietnam 134
Volkswagen 75, 76, 115
Volvo 75, 76, 115
vulnerability 22, 133
vulnerable populations, resilient and inclusive focused on 7–9

Wallace, G.C. 203n1
Wall Street Journal 55
Walmart 114, 200; contradictions with business goals 121
Walt Disney Company, The 55
Walter Energy 31
WEF *see* World Economic Forum (WEF)
Wellcome 188
Wells Fargo: contradictions with business goals 121
We Mean Business Coalition 111, 136
WFP *see* World Food Programme (WFP)
WHO *see* World Health Organization (WHO)
Williams–Sonoma 117
WMO *see* World Meteorological Organization (WMO)
women empowerment, strengthening 141–144
Woodside Energy 201
workplace safety 27–28
World Bank 103, 104, 186

World Economic Forum (WEF) 1, 11, 26, 122; *Davos Manifesto 2020: the Universal Purpose of a Company in the Fourth Industrial Revolution, The* 54
World Food Programme (WFP): on acute food insecurity 4
World Health Organization (WHO) 73
World Meteorological Organization (WMO) 5
World Resources Institute (WRI) 7, 95
Worldwide Fund for Nature 189

WRI *see* World Resources Institute (WRI)
WWF (formerly the World Wildlife Fund) 117, 189

XR *see* Extinction Rebellion (XR)

Yale's School of Management 38
Yale University 136

Zucman, G. 146

Made in the USA
Columbia, SC
06 January 2022